国家重大出版工程项目

犬猫肥胖症与管理策略

Obesity in the Dog and Cat

Martha G. Cline　Maryanne Murphy　主编

孙伟丽　刘可园　李　平　主译

中国农业大学出版社
·北京·

内 容 简 介

目前,在全球范围内,有许多超重及肥胖的犬和猫,肥胖导致其生活质量下降,寿命缩短,患其他疾病的风险增加,进而导致兽医护理费用整体大幅度增加。本书内容包含犬猫肥胖症的主要原因、临床治疗案例、兽医师执业经验,并给出了针对肥胖犬猫减肥的营养餐食谱,更重要的是本书提供了一种宠物健康营养理念。本书集结了许多关于犬猫肥胖诊疗的案例资源,可供第一线的兽医专业人员参考,也希望本书能给予对犬猫营养感兴趣的人一定的启发。

图书在版编目(CIP)数据

犬猫肥胖症与管理策略/(美)玛莎·克莱因(Martha G. Cline),(美)玛丽安妮·墨菲(Maryanne Murphy)主编;孙伟丽,刘可园,李平主译. --北京:中国农业大学出版社,2022.10

书名原文:Obesity in the Dog and Cat

ISBN 978-7-5655-2852-1

Ⅰ.①犬… Ⅱ.①玛… ②玛… ③孙… ④刘… ⑤李… Ⅲ.①犬病-肥胖病 ②猫病-肥胖病 Ⅳ.①S858.292 ②S858.293

中国版本图书馆 CIP 数据核字(2022)第 140675 号

书　　名	犬猫肥胖症与管理策略		
	Quanmao Feipangzheng yu Guanli Celüe		
作　　者	Martha G. Cline　Maryanne Murphy　主编		
	孙伟丽　刘可园　李　平　主译		
策划编辑	梁爱荣	责任编辑	胡晓蕾
封面设计	郑　川		
出版发行	中国农业大学出版社		
社　　址	北京市海淀区圆明园西路 2 号	邮政编码	100193
电　　话	发行部 010-62733489,1190	读者服务部	010-62732336
	编辑部 010-62732617,2618	出 版 部	010-62733440
网　　址	http://www.caupress.cn	E-mail	cbsszs@cau.edu.cn
经　　销	新华书店		
印　　刷	涿州市星河印刷有限公司		
版　　次	2022 年 10 月第 1 版　2022 年 10 月第 1 次印刷		
规　　格	170 mm×240 mm　16 开本　13.5 印张　255 千字		
定　　价	120.00 元		

图书如有质量问题本社发行部负责调换

Obesity in the Dog and Cat / by Martha G. Cline, Maryanne Murphy / ISBN: 978-1-4987-4147-7

Copyright© 2019 by Taylor & Francis Group, LLC.

Authorized translation from English language edition published by 2019 by Taylor & Francis Group, LLC; All rights reserved; 本书原版由 Taylor & Francis 出版集团出版，并经其授权翻译出版. 版权所有，侵权必究.

China Agricultural University Press Ltd. is authorized to publish and distribute exclusively the Chinese (Simplified Characters) language edition. This edition is authorized for sale throughout Mainland of China. No part of the publication may be reproduced or distributed by any means, or stored in a database or retrieval system, without the prior written permission of the publisher. 本书中文简体翻译版授权由中国农业大学出版社有限公司独家出版并限在中国大陆地区销售。未经出版者书面许可，不得以任何方式复制或发行本书的任何部分.

Copies of this book sold without a Taylor & Francis sticker on the cover are unauthorized and illegal. 本书封面贴有 Taylor & Francis 公司防伪标签，无标签者不得销售.

著作权合同登记图字：01-2022-4844

本书的翻译出版得到下列机构的资助：
天津朗诺宠物食品股份有限公司
北京奥特奇生物制品有限公司
上海依蕴宠物用品有限公司

本书的翻译还得到中国农业科学院特产研究所、青岛农业大学和广东省农业科学院动物科学研究所的支持和帮助。

在此特别致谢！

译 者 名 单

主译 孙伟丽 青岛农业大学/中国农业科学院特产研究所
　　　　刘可园 青岛农业大学/中国农业科学院特产研究所
　　　　李　平 广东省农业科学院动物科学研究所

参译（按姓氏笔画排序）
　　　　王　丁 芜湖海派宠物用品有限公司
　　　　王　卓 中国农业科学院特产研究所
　　　　付曦瑶 中国农业科学院在读研究生
　　　　刘凤华 山东畜牧兽医职业学院宠物科技学院
　　　　刘文峰 广州优百特科技有限公司/
　　　　　　　　沈阳优佰特生物科技有限公司
　　　　孙皓然 中国农业科学院在读博士
　　　　张　婷 中国农业科学院特产研究所
　　　　张博睿 青岛农业大学在读研究生
　　　　赵梦迪 吉林农业大学在读博士
　　　　郭颖妍 赤峰应用技术职业学院
　　　　梁晓欢 华南农业大学兽医学院

撰 稿 者

Charlotte Reinhard Bjørnvad，兽医师，博士，毕业于欧洲兽医与比较营养学院，任职于丹麦哥本哈根市哥本哈根大学医学、肿瘤学和兽医临床病理学部。

Lene Elisabeth Buelund，兽医师，博士，任职于丹麦哥本哈根市哥本哈根大学兽医临床系兽医影像科。

Martha G. Cline，兽医师，营养学专家，任职于美国新泽西州廷顿瀑布市红岸动物医院。

Ashley Cox，理学学士，兽医技术专家（营养学领域），任职于美国田纳西州诺克斯维尔市田纳西大学兽医学院小动物临床科学系。

Alexander J. German，兽医学学士，博士，获得内科专科认证证书（CertSAM），是欧洲伴侣动物内科协会（DECVIM-CA）专家，英国高等教育学会（SFHEA）成员，英国皇家兽医学院（FRCVS）院士，任职于英国内斯顿利物浦大学健康与生命科学学院兽医科学研究所，衰老与慢性疾病研究所。

Kenneth J. Lambrecht，兽医师，任职于美国威斯康星州风险体重管理系统和麦迪逊西镇兽医中心。

Deborah E. Linder，兽医师，营养学专家，任职于美国马萨诸塞州北格拉夫顿塔夫茨大学塔夫茨人与动物关系研究所卡明斯兽医学院临床科学系。

Lydia Love，兽医师，麻醉和镇痛专业（DACVAA），任职于美国新泽西州费尔菲尔德动物危重症和转诊协会。

Andrew McGlinchey，兽医师，任职于美国加利福尼亚州洛杉矶市洛杉矶VCA医院。

Megan K. Mueller，文科硕士，博士，任职于美国马萨诸塞州北格拉夫顿塔夫茨大学卡明斯兽医学院临床科学系，乔纳森·蒂施公民与公共服务学院。

Maryanne Murphy，兽医师，博士，营养学专家，任职于美国田纳西州诺克斯维尔市田纳西大学兽医学院小动物临床科学系。

Elizabeth Orcutt，兽医师，理学硕士学位，心脏病学专家，任职于美国新泽西州廷顿瀑布市红岸动物医院。

Valerie J. Parker,兽医师,毕业于美国兽医学院,兽医内科学、营养学,任职于美国俄亥俄州哥伦布市俄亥俄州立大学兽医学院兽医临床科学系。

Angela Witzel Rollins,兽医师,博士,营养学专家,任职于美国田纳西州诺克斯维尔市田纳西大学兽医学院小动物临床科学系。

Megan Shepherd,兽医师,博士,营养学专家,任职于美国弗吉尼亚州布莱克斯堡镇弗吉尼亚-马里兰兽医学院大型动物临床科学系。

Justin Shmalberg,兽医师,营养学专家,任职于美国佛罗里达州盖恩斯维尔市佛罗里达大学兽医学院诊断比较和医学人口系。

Moran Tal,兽医师,理学学士,兽医学博士,任职于加拿大安大略省圭尔夫市圭尔夫大学安大略兽医学院临床研究系。

Adronie Verbrugghe,兽医师,博士,毕业于欧洲兽医与比较营养学院,任职于加拿大安大略省圭尔夫市圭尔夫大学安大略兽医学院临床研究系。

Claudia Wong,兽医师,任职于加拿大安大略省圭尔夫市圭尔夫大学安大略兽医学院临床研究系。

序

　　截至2021年,中国宠物犬猫数量已经超过1亿只,宠物健康是宠物主人关注的首要问题。其中,由于缺乏肥胖症的相关知识经验、不科学的喂养习惯、过度饲喂和不均衡饲喂、体况和运动管理不足等原因,都可能导致犬猫肥胖症的发生和加重。相信阅读本书将有助于读者了解犬猫肥胖症的原理和应对策略,注重犬猫的日常营养需要和体态管理,培养健康科学的喂养和运动习惯。

　　犬猫肥胖症的治疗是一个典型的临床营养问题,需要兽医师精通犬猫营养学,尤其是肥胖相关营养学的知识。我国对犬猫肥胖症的研究还处于起步阶段,需要相关科研人员和从业人员加强基础学习和临床实践。本书从犬猫肥胖症的流行病学、病理生理学,以及犬猫肥胖症的临床诊断、营养管理、行为管理和临床案例等多个层面,系统地介绍了犬猫肥胖症的理论基础和临床治疗方案,既有理论指导,又有实践案例,对犬猫肥胖症的预防和治疗具有很高的参考价值。

　　本译著的翻译团队由国内高校及研究院所十余位硕博士组成,他们从事动物营养及动物医学等领域科研、教学及公司技术指导等工作,具有严谨的科学态度,扎实的理论基础,翻译忠实于原作者想表达的思想。希望通过本书,能给我国宠物营养领域带来新知!让更多宠物更健康!未来希望有更多同人关注宠物营养领域并能不断推动好书的翻译出版!

青岛农业大学动物科技学院教授　李光玉
2022年5月

中文版前言

宠物肥胖问题日益普遍和突出，欧美等发达国家约 50% 的宠物犬猫存在肥胖问题，这对犬猫健康造成一系列危害。近些年，国内的宠物主人也越来越关注宠物肥胖的问题。然而，国内专门针对犬猫肥胖症并可为宠物主人提供系统全面知识的专业书籍却极为稀缺。由此，我们从 20 多本备选书目中，选择了这本《犬猫肥胖症与管理策略》进行翻译出版。

这本专业著作非常务实精炼，架构科学，系统全面，图文表并茂，不到 200 页的篇幅却囊括了犬猫肥胖症的流行病学特点、病理生理学原理、合并症和麻醉、诊断、营养和行为管理策略、运动作用、临床诊疗的体重管理和病例分析等内容，看似是一本适合普通宠物主人阅读的科普书，却也是一本非常优秀、适合专业人士参考学习的专业书。

本书的两位主编 Martha G. Cline 和 Maryanne Murphy 在小动物临床营养领域具有多年临床诊治和基础科研经验，并参与制定了美国动物医院协会(the American Animal Hospital Association, AAHA)犬猫营养与体重管理指南，是美国业内的权威专家。

为了更好地完成这本专著的翻译工作，我们邀请了动物医学和动物营养学的专家和学者，也邀请了国内外宠物相关企业的技术专家作为翻译团队的成员，在此特别致谢！另外，还特别感谢上海依蕴宠物用品有限公司的赵海明先生和迟春艳女士，上海艾尼贸沃生物科技有限公司的王廷廷女士，北京奥特奇生物制品有限公司的刘婷婷女士，以及天津朗诺宠物食品股份有限公司刘凤岐先生的帮助！

由于译者水平有限，翻译不当之处欢迎各位读者批评指正。

译　者
2022 年 8 月

英文版前言

兽医是一个需要面对严重流行病的职业。犬猫超重和肥胖症会导致宠物的生活质量下降、寿命缩短、增加其患其他疾病的风险，并提高整体医疗成本。兽医专业人员，从护士到专业委员会的专家，都应该熟练掌握宠物肥胖症这种流行病的特征和规律。

本书适用于全科兽医和兽医护士。我们供职于一家大型私人宠物医院，在这家宠物医院开始了职业生涯，并与兽医专家和全科兽医一起参与宠物诊疗。尽管市面上可见到许多关于宠物肥胖症的优质资源和参考资料，但我们认为一线兽医专业人员仍然需要一本全面阐述宠物肥胖症及管理策略的书籍。我们也希望本书所总结的最新研究成果，能给那些对兽医营养学有浓厚兴趣的读者带来一定的启发。

我们要感谢本书作者们的贡献。他们在临床诊疗中解决了宠物肥胖症相关的诸多问题。他们通过科研试验和发表论文，为我们提供了有关宠物肥胖症方面的知识。在兽医学肥胖症领域，他们为即将从事和目前已从事兽医专业的人员，提供了全方位的知识。这些专家丰富的经验和知识，将会使本书成为兽医专业人员的有益参考书。

Martha G. Cline
Maryanne Murphy

致　谢

我们想把这本书献给亲爱 B. J. 和 Cormac,以感谢他们无限的耐心和支持。
我们也把这本书献给 Colm,Clara,Ryder,Lennert,Jaxen 和 Yona。

目 录

1 小动物肥胖症的流行病学特点 ·· 1
  1.1 定义 ··· 1
  1.2 患病率和时间趋势 ··· 1
  1.3 风险因素 ··· 4
    1.3.1 动物自身因素 ··· 4
    1.3.2 人为因素 ··· 6
  1.4 结论 ··· 8
  参考文献 ··· 8
2 肥胖症的病理生理学:代谢效应和炎症介质 ····················· 14
  2.1 能量消耗与能量摄入的平衡 ·································· 14
    2.1.1 肥胖症的进化理论 ······································ 14
    2.1.2 食欲和采食调节 ··· 16
    2.1.3 能量消耗和能量代谢调控机制 ····················· 18
  2.2 肥胖症的炎症效应 ·· 19
    2.2.1 脂肪组织的内分泌功能 ································ 19
    2.2.2 肥胖症和炎症 ·· 20
  2.3 微生物在肥胖症中的作用 ···································· 21
    2.3.1 肠道微生物菌群与肥胖症 ···························· 21
    2.3.2 肠道微生物菌群在肥胖症治疗中的应用 ········ 22
  2.4 结论 ·· 22
  参考文献 ·· 23
3 肥胖症的病理生理学:合并症和麻醉的注意事项 ············· 33
  3.1 引言 ·· 33
  3.2 肥胖症和寿命 ·· 33
  3.3 肥胖症悖论 ··· 34

I

3.4 内分泌疾病 35
3.4.1 糖尿病 35
3.4.2 犬糖尿病 35
3.4.3 猫糖尿病 36
3.4.4 肾上腺皮质功能亢进 37
3.4.5 甲状腺功能减退 37
3.4.6 高脂血症 37

3.5 心血管和呼吸系统疾病 38
3.5.1 充血性心力衰竭和心血管疾病 38
3.5.2 高血压 39
3.5.3 气管塌陷 39

3.6 肾脏和泌尿疾病 39
3.6.1 肾脏疾病 39
3.6.2 泌尿疾病 40

3.7 骨科和神经内科疾病 41
3.7.1 骨科疾病 41
3.7.2 神经内科疾病 42

3.8 肿瘤 42

3.9 麻醉注意事项 43
3.9.1 肥胖症引起的生理变化 43
3.9.2 肥胖症所致的药理学差异 45

3.10 结论 46

参考文献 46

4 通过评价体成分诊断宠物肥胖症 55
4.1 引言 55
4.2 体成分的胴体分析 55
4.3 影像学方法评估体成分 57
4.3.1 双能X射线吸收仪 57
4.3.2 计算机断层扫描 58
4.3.3 核磁共振成像 59
4.4 非影像学方法评估体成分 60
4.4.1 重水稀释技术 60
4.4.2 生物电阻抗 61

4.5 临床环境下的体成分评估 ··· 62
4.5.1 体重 ··· 62
4.5.2 形态测量 ··· 62
4.5.3 形态学估算 ··· 63
4.6 全科诊疗的临床建议 ··· 65
4.7 总结 ··· 65
参考文献 ··· 66

5 肥胖症宠物的营养管理策略 73
5.1 肥胖宠物评估 ··· 73
5.1.1 宠物肥胖症的诊断 ··· 73
5.1.2 估算目标体重或理想体重 ··· 77
5.2 体重管理的能量需要 ··· 78
5.2.1 确定当前热量的摄入量 ··· 78
5.2.2 确定减肥的能量摄入量 ··· 81
5.3 实施治疗方案 ··· 84
5.3.1 食物选择 ··· 84
5.3.2 减肥食物分配 ··· 87
5.3.3 制定饮食指南 ··· 87
5.4 重新评估减肥计划 ··· 88
5.4.1 例行随访 ··· 88
5.4.2 保持长期成功效果 ··· 88
5.5 结论 ··· 89
参考文献 ··· 89

6 肥胖症宠物的行为管理策略 94
6.1 引言 ··· 94
6.2 宠物主人主观认知对宠物行为和肥胖症的影响 ··· 94
6.2.1 对宠物肥胖症的认知 ··· 95
6.2.2 了解宠物主人的想法 ··· 95
6.3 将宠物主人心理和行为的管理纳入宠物肥胖症治疗方案 ··· 97
6.3.1 与宠物主人讨论宠物的肥胖问题 ··· 97
6.3.2 在最初的体重讨论中参考兽医的意见 ··· 99
6.4 通过解决情绪行为来制定有效的体重管理计划 ··· 99
6.5 考虑家庭成员互动在管理肥胖行为中的作用 ··· 100
6.6 宠物与主人共同减肥的互动行为 ··· 101

	6.7	结论		102
	参考文献		102
7	运动在宠物肥胖症治疗中的作用		105
	7.1	引言		105
	7.2	运动与肥胖症风险		105
	7.3	运动治疗人类肥胖症		106
	7.4	犬和猫合并症、运动和肥胖症		107
	7.5	运动在犬和猫肥胖症治疗中的应用		107
	7.6	实用建议		109
	7.7	结论		110
	参考文献		110
8	临床诊疗中建立体重管理程序		115
	8.1	引言		115
	8.2	初次会诊		115
		8.2.1	日粮组成、活动情况和家庭生活史	115
		8.2.2	把握初次会诊的信息	116
	8.3	随访和复查预约		118
	8.4	维持期的复诊		120
	8.5	宣传营销		122
	8.6	经济因素的影响		126
	8.7	结论		128
	附录 8.1	饮食、活动和家庭生活史记录表（DAHHF）	129
	附录 8.2	客户照片发布许可表	136
	附录 8.3	饮食限额表	136
	附录 8.4	每周饮食日记	139
	参考文献		139
9	病例分析		142
	9.1	病例 1：一只腊肠犬椎间盘突出手术后成功减肥并维持体重——风险因素评估的重要性	142
		9.1.1	既往病史	142
		9.1.2	病情评估	142
		9.1.3	治疗方案	143
		9.1.4	短期和长期随访	143
		9.1.5	讨论	144

参考文献	145

9.2 病例2:一只成年绝育雌性史宾格猎犬的病态肥胖症 …………… 147
- 9.2.1 既往病史 …………… 147
- 9.2.2 病情评估 …………… 148
- 9.2.3 减肥计划 …………… 148
- 9.2.4 定期复查 …………… 149
- 9.2.5 讨论 …………… 150

参考文献	151

9.3 病例3:肥胖小动物的麻醉问题 …………… 151
- 9.3.1 既往病史 …………… 151
- 9.3.2 患猫评估 …………… 151
- 9.3.3 治疗方案 …………… 151
- 9.3.4 短期和长期随访 …………… 153
- 9.3.5 讨论 …………… 153

参考文献	155

9.4 病例4:减肥计划中肥胖的短毛猫肥胖程度估算 …………… 156
- 9.4.1 既往病史 …………… 156
- 9.4.2 病情评估 …………… 156
- 9.4.3 治疗方案 …………… 156
- 9.4.4 短期随访 …………… 157
- 9.4.5 讨论 …………… 158

参考文献	159

9.5 病例5:家养短毛猫减肥对糖尿病的缓解作用 …………… 160
- 9.5.1 既往病史 …………… 160
- 9.5.2 患猫评估 …………… 160
- 9.5.3 治疗方案 …………… 160
- 9.5.4 短期和长期随访 …………… 161
- 9.5.5 讨论 …………… 162

参考文献	163

9.6 病例6:肥胖症的行为管理 …………… 164
- 9.6.1 患犬病史 …………… 164
- 9.6.2 患犬评估 …………… 164
- 9.6.3 治疗方案 …………… 164
- 9.6.4 短期和长期随访 …………… 165

    9.6.5 讨论 ………………………………………………………… 165
  参考文献 ………………………………………………………………… 166
  9.7 案例 7：一例新型蛋白质饮食试验中的热量限制、运动和
    治疗性减肥 …………………………………………………………… 166
    9.7.1 既往病史 ……………………………………………………… 166
    9.7.2 患犬评估 ……………………………………………………… 167
    9.7.3 治疗方案 ……………………………………………………… 168
    9.7.4 随访 …………………………………………………………… 170
    9.7.5 讨论 …………………………………………………………… 171
  参考文献 ………………………………………………………………… 172
  9.8 病例 8：体重减轻计划中兽医护士的作用 ………………………… 174
    9.8.1 既往病史 ……………………………………………………… 174
    9.8.2 饮食史 ………………………………………………………… 174
    9.8.3 患犬评估 ……………………………………………………… 174
    9.8.4 治疗方案 ……………………………………………………… 175
    9.8.5 随访 …………………………………………………………… 176
    9.8.6 讨论 …………………………………………………………… 177
  参考文献 ………………………………………………………………… 179
索引 …………………………………………………………………………… 180

1 小动物肥胖症的流行病学特点

1.1 定义

肥胖是由于长期的能量正平衡造成的,当能量摄入超过能量需要时,体内脂肪的沉积导致体重超标[1-3]。当宠物体重比理想体重高 10%~20% 时,视为超重;当体重比理想体重高 20% 以上时视为肥胖[4]。宠物体成分方面的研究,将体脂肪含量在 25%~35% 定义为超重,体脂肪含量大于 35% 定义为肥胖[5,6]。由于过多的体脂肪会改变身体的功能并诱导疾病发生,因此这些与肥胖症有关的定义也应该包含一部分疾病方面的考虑[1-3]。

1.2 患病率和时间趋势

肥胖症是伴侣动物面临的头号营养问题。据估计,犬类肥胖症在工业化国家的患病率为 11.2%~59.4%(表 1.1)[7-20],猫肥胖症的患病率为 11.5%~63.0%(表 1.2)[10, 21-32]。

表 1.1　55 年间(1960—2015 年)各国犬肥胖症的患病率

作者	年份	国家	样本群体来源	样本量	评价方法	患病率
Krook 等[7]	1960	瑞典	犬尸体剖检	10 993	病理医生 病理性肥胖	肥胖:11.2%
Mason[8]	1970	英国	1 家宠物医院	1 000	主治兽医 3 分制:瘦、正常、肥胖, 脂肪覆盖胸部	肥胖:28%
Edney 和 Smith[9]	1986	英国	11 家宠物医院, 范围从兽医学校 到私人宠物医院	8 268	主治兽医 5 分制:瘦、偏瘦、最佳、 肥胖、过度肥胖	肥胖:24.3%
Lund 等[10]	1999	美国	52 家宠物医院	86 772	主治兽医 5 分制:脂肪覆盖肋骨、 尾根部和腹部轮廓	BCS ≥ 4/5: 28.3%

续表 1.1

作者	年份	国家	样本群体来源	样本量	评价方法	患病率
Robertson[11]	2003	澳大利亚	2 326 只家养犬，电话调查	657	宠物主人 3 分制：体重不足、标准体重、超重/肥胖	超重/肥胖：25.2%
McGreevy 等[12]	2005	澳大利亚	209 家宠物医院	2 661	主治兽医 5 分制：根据希尔(Hill)犬猫体型示意图进行评分	BCS ≥ 4/5：1.1%
Colliard 等[13]	2006	法国	兽医学校的疫苗接种服务	616	主治兽医 5 分制，改编自 Laflamme[5]	BCS ≥ 4/5：38.8%
Lund[14]	2006	美国	52 家宠物医院	21 754	主治兽医 5 分制：脂肪覆盖肋骨、尾根部和腹部轮廓	BCS ≥ 4/5：34.1%
Courcier 等[15]	2010	英国	1 家慈善宠物医院和 4 家私人初诊(全科)宠物医院	696	主治兽医或兽医实习生体况评分方法改编自 S.H.A.P.E.™ 的 7 分制形态测量方法[74]	S.H.A.P.E.™ ≥ 5/7：59.4%
Sallander 等[16]	2010	瑞典	纯种犬的主人，在大型动物保险公司注册的犬，电话调查	461	宠物主人 5 分制：非常瘦、瘦、正常、肥胖、非常肥胖[75,76] (采用 Laflamme[5] 和过去的评分示意图)	肥胖/过度肥胖：16%
Heuberger 和 Wakshlag[17]	2011	美国	采用口头宣传和广告招募的犬主人，邮件调查	61	宠物主人 5 分制：体重不足、轻微体重不足、理想体重、轻微超重、肥胖	轻微超重/肥胖：21.3%
Mao 等[18]	2013	中国	14 家宠物医院	2 391	主治兽医 5 分制[4,5]	BCS ≥ 4/5：44.4%
Corbee[19]	2013	荷兰	1 场犬展	1 379	专业评委会认证的兽医营养师™ 9 分制[5]	BCS ≥ 6/9：19.8%
Such 和 German[20]	2015	英国	在犬展中参展犬的图片	960	有经验的兽医从照片上进行体况评分(BCS) 使用过去 9 分制和过去的评分示意图	BCS ≥ 6/9：26.0%

注：BCS——体况评分；S.H.A.P.E.™——体型、健康和身体评估。

表 1.2　23 年间(1992—2015 年)各国猫肥胖症的患病率

作者	年份	国家	样本群体来源	样本量	评价方法	患病率
Sloth[21]	1992	丹麦	1 家宠物医院	233	主治兽医 4 分制:体重不足、正常体重、超重、肥胖	超重/肥胖:40%
Scarlett 等[22]	1994	美国	31 家宠物医院	2 091	主治兽医 6 分制(根据体型划分):极瘦弱、消瘦、理想偏瘦、理想体重、偏重、肥胖	偏重/肥胖:25%
Donoghue 和 Scarlett[23]	1998	美国	27 家宠物医院,来自之前对猫的研究资料[22]	1 654	主治兽医 6 分制[22]	偏重/肥胖:24.5%
Lund 等[10]	1999	美国	52 家宠物医院	42 774	主治兽医 5 分制:脂肪覆盖肋骨、尾根部和腹部轮廓,改编自 Scarlett 等的方法[22]	BCS ≥ 4/5:27.5%
Robertson[24]	1999	澳大利亚	2 195 只养猫,电话调查	644	宠物主人 3 分制:体重不足、正常体重、超重(肥胖)	超重/肥胖:18.9%
Russell 等[25]	2000	英国	养猫家庭,上门拜访	136	有经验的评估人员 改编自 Laflamme[6] 制定的 17 分制	超重/肥胖:52%
Allan 等[26]	2000	新西兰	492 只家养猫,上门调查	182	有经验的评估人员 3 分制(评估腹股沟、腹部和皮下脂肪):正常体重、超重、肥胖	超重/肥胖:25.8%
Lund 等[27]	2005	美国	52 家宠物医院	8 159	主治兽医 5 分制:脂肪覆盖肋骨和腹部轮廓	BCS ≥ 4/5:35.1%
Colliard 等[28]	2009	法国	兽医学校的疫苗接种服务	385	主治兽医 改编自 Laflamme[6] 制定的 5 分制	BCS ≥ 4/5:26.8%
Courcier 等[29]	2010	英国	1 家慈善初诊(全科)宠物医院	118	兽医学生 改编自 Laflamme[6] 制定的 5 分制	BCS ≥ 4/5:39.0%

续表 1.2

作者	年份	国家	样本群体来源	样本量	评价方法	患病率
Courcier 等 [30]	2012	英国	47家慈善初诊（全科）宠物医院	3 219	主治兽医 5分制	BCS ≥ 4/5：11.5%
Cave 等 [31]	2012	新西兰	1 045只家养猫，上门调查，与Allen等[26]相似	200	2位独立的、有经验的评估人员 9分制[6]	BCS ≥ 7/9[a]：27.4% BCS ≥ 6/9：63.0%
Corbee [32]	2014	荷兰	2次猫展	268	专业评委会认证的兽医营养师™ 9分制[6]	BCS ≥ 6/9：45.5%

注：BCS：体况评分。
[a] 可以将经过验证的9分制[6]简化为未经验证的5分制，其中：1分=1分，2~3分=2分，4~6分=3分，7~8分=5分，9分=5分[31]。

然而，这些数据有很多局限性。有的研究时间跨度很大，可能无法完全反映当前的情况。而且很多研究是基于数量有限的宠物医院案例得出的结果，或者是在特定的宠物群体中开展的研究，且使用不同方法评定宠物的体况也会导致对于超重/肥胖的定义有所差异。此外，不同研究目的和研究方向的科研人员对宠物体况的定义也有所不同。

因此，很难调查清楚宠物肥胖症发病率在不同国家之间的差异和时间趋势。在一项为期4年的调查中，猫群体的体况分布相对稳定[22,23]。同样，在1993年和2007年，对同一城市养猫家庭的调查中也没有发现肥胖症患病率存在差异[26,31]。

1.3　风险因素

肥胖症是由多种因素引起的，许多风险因素都可影响能量摄入和利用，造成能量正平衡[1,2]。兽医们认为，只有3%的肥胖病例是由动物自身因素引起的，而97%是由人为因素引起，如饮食因素、运动量、主人态度和家庭特征（图1.1）[33]。

1.3.1　动物自身因素

1.3.1.1　遗传和品种因素

动物肥胖症的遗传因素可以解释为品种易感性（breed predisposition）。尽管如此，对某些犬品种肥胖症的研究会部分取决于某一区域主要养哪个品种。总的来说，巴吉特猎犬、小猎犬、凯恩㹴、可卡犬、达克斯猎犬、金毛猎犬、拉布拉多猎犬、哈巴犬和设得兰牧羊犬都属于易患肥胖症的犬种[8,9,13,14,18-20]。原因可能是这些品种

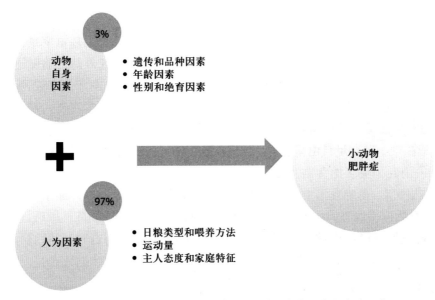

图1.1 小动物肥胖症的风险因素，包括动物自身因素和人为因素
（数据引自 Bland 等报道[33]）

出于特定目的选育而成或其体质状况不同。对于某些犬种，育种标准就是向超重选育的[19,20]。此外，不同品种之间的体成分不同[34]，也会影响能量消耗[35]。研究表明杂交猫超重的风险更高[22,24,27,28]。在纯种猫中，英国短毛猫、波斯猫和挪威森林猫的体况评分比较高[32,36]。

1.3.1.2 年龄因素

肥胖症患病率随着年龄的增长而增加[8,11-13,18,22,31]，这是由于随年龄增长能量需要逐渐减少，特别是随着运动量的减少和与年龄相关的体成分的变化，导致宠物的基础代谢率下降[37,38]。Harper 的研究表明，犬成年后的能量需要减少了20%，而猫在成年期能量需要是恒定的[37]。但是，Bermingham 等的研究未能证实老年犬的能量需要有所下降[39]，却发现猫的能量消耗与年龄有关[40]。然而，流行病学研究表明，犬[12,14,18]和猫[22,24,27,28,30]都是在中年时肥胖症患病率最高。犬猫肥胖症患病率大约在10岁之后开始下降[12,22]，原因可能是因为超重有损健康，从而导致寿命缩短[41]。

1.3.1.3 性别和绝育因素

流行病学研究显示母犬的超重比例较高[7,8,12,13,15,16,18]，这可能是由于母犬的体成分和能量需要与公犬有差异。与母犬相比，公犬身体含有更多的瘦肉组织[42]，这意味着母犬的能量需要更低，容易导致体重增加。然而，最近的一项大数据分析并未证实性别会影响犬的能量需要[39]。在对猫的研究中发现，母猫虽然比公猫的能量需要更低[43],但公猫却更容易超重[22,24,27,28,30,36]。

绝育与肥胖症密切相关[9,11,13,14,18,21,22,24,27-30,32,36]。绝育犬的超重率是正常犬的2.8倍[11],而绝育猫的超重率是正常猫的3.6倍[30]。与5.5月龄之前做绝育相比,5.5~12月龄做绝育的犬出现肥胖症的概率更高[44]。然而,对于犬来说,小月龄做绝育手术发生肥胖症的概率更低,这种现象在猫中还未得到证实[45]。肥胖症与绝育的相关性是由性激素减少导致。雌性动物可能会出现由雌激素诱导的食欲抑制现象[46]。当充足自由采食时,与未绝育的对照组相比,绝育后的雌性比格犬(Beagle)消耗了更多的食物,增加了更多的体重[46]。与未绝育的对照组或与绝育前的猫相比时,在绝育猫中也发现了相似的结果[47-49]。因此,为了使母猫[49,50]和母犬绝育后保持理想的体重,必须严格限制食物摄入量。猫在切除性腺后,无论公母,其静息能量消耗均降低了30%[52,53]。最后,由于减少了闲逛时间和性行为,绝育还会导致自发活动减少,雄性动物表现得更明显[54,55]。

1.3.2 人为因素

1.3.2.1 日粮类型和喂食方式

美味可口、富含能量的食物会让宠物面临体重增加的风险,尤其是在自由选择食物的情况下。日粮中的脂肪是最富集和最有效的能量来源[3]。在犬[56]和猫[57,58]中,相比于低脂肪、高碳水化合物的饮食,高脂肪、低碳水化合物日粮更容易导致体脂和体重增加。尽管如此,所有常量营养素(脂肪、蛋白质和碳水化合物)都是以体脂的形式储存,如果摄入过量就会导致体重增加[3]。相反,日粮中加入膳食纤维、水或空气可降低能量密度。在犬中,高膳食纤维含量的日粮与正常体况之间存在正相关关系[17]。

大多数研究发现商业宠物食品类型对肥胖症患病率没有影响[9,11,15,24,26,29,31]。尽管如此,一些报告指出,猫食用"优质"干粮和/或治疗性日粮患病的风险更高[22,27,28],这可能是因为"优质"干粮和/或治疗性日粮比大多数经济型日粮能量密度更高[22,27]。此外,宠物主人更倾向于让宠物随意采食干粮[23]。尽管宠物主人喂食干粮的比例有所增加,但澳大利亚研究人员在1993年[26]和2007年[31]的研究表明,肥胖症与喂食干粮没有关系,在研究期间肥胖症也没有增加[31]。而且,流行病学的研究也无法确定治疗性日粮与肥胖症的因果关系。值得注意的是,当低能量治疗性日粮用于减肥时,可能会使这一关系发生混淆[27,28]。同时有其他研究显示,喂食家庭自制食物、餐桌剩饭和/或零食的犬和猫更容易患肥胖症[8,11,14-18,59,60]。

自由选择或随意喂食会导致能量摄入过量,而将每天的食物分成多餐可以增加由产热[61]和身体活动造成的能量消耗[62]。许多研究对随意喂食和喂食频率影响存在分歧,而且两者往往难以区分。在不同的研究中,犬每天进食1次[11]、多次[18],

以及喂食1份或3份以上[59]时,患肥胖症的风险均有升高。在猫上的研究也并不一致,一些研究认为随意喂食或增加喂食频率是导致肥胖的危险因素[25,48,60],但是另一些研究则不这样认为[22,24,26,28,31],还有一项研究发现,每天喂食2次的猫比随意喂食的猫更容易肥胖。虽然宠物主人每天可能只给宠物1次或2次食物,但食物的总量却可能满足了全天需要[24]。此外,将每天适量的食物分成小份喂食,或者通过自由选择食物而提供了多余的能量,这2种方式也不应该混淆。

总之,过度喂食更能导致宠物肥胖症。由于产品推荐量是基于能量需要的范围和平均值,因此根据包装说明书确定每日摄入量可能会导致能量摄入过多。当宠物主人高估了他们宠物的运动量[63]并且低估了宠物体况的情况下[13,26,29,31,64-66],过度喂食的可能性就大大增加。70%的猫主人会一直喂食,直到他们的猫不吃了为止,仅有8%的人会按照包装上的说明去喂食,仅有4%的人会询问他们的宠物医生[29]。此外,大多数宠物主人不会去使用可精确至克的秤称量食物,而是用量杯。使用量杯虽然方便快捷,但不精确也不准确,这也是导致肥胖症流行的因素之一[67]。

1.3.2.2 运动量

缺乏运动量也是导致肥胖症的一个主要危险因素[11,15,18,21,22,24,26,59],运动量不足会减少能量消耗[68]。大多数研究将其归因于宠物的生活环境类型。肥胖的宠物往往生活在室内,运动可能受到限制[11,21,22,24,26]。尽管如此,2项关于猫的研究并不足以将运动与肥胖症联系起来,这可能是由于不同国家对宠物活动限制条件有所不同,也可能是由于宠物主人对宠物运动量的评估存在差异[25,31]。然而,宠物生活环境可能不是最好的评估指标。只能在院子里活动的犬比去户外散步的犬更容易肥胖[59],住在公寓中的犬比住在独栋房子的犬活动更加频繁[69]。总的来说,犬类患肥胖症的风险会随着每周活动时间的增加而降低[11,15],但是运动强度(散步或者跑步)对肥胖症没有影响[11]。遗憾的是,大多数宠物主人高估了犬的运动量,他们认为自己的爱犬运动量适度甚至非常活跃[63]。同时,运动量少的宠物,其主人往往运动量也不高[70]。

1.3.2.3 宠物主人态度和家庭特征

通常,肥胖的宠物被其主人过度人性化,并且可能替代人类成为其主人的陪伴者。例如,肥胖的犬经常睡在主人的床上,主人与他们的犬和猫交谈得更多[60,71]。这些宠物不再被视为一般的伴侣动物,往往淡化了其在诸如运动、玩耍、为人类服务、警卫和丰富生活氛围等方面的作用[60,71]。因此宠物的健康和营养需要也被忽略了[60,71]。宠物的行为信号经常被"误解",导致主人以为宠物饿了从而过度喂食,而实际上它并不饿,而且食物也经常用来作为物质奖励[71]。超重犬的主人很少关注犬粮的营养平衡,他们更关注价格和特惠活动[71,72],而不是原料的质量和营养成分[71,72]。部分原因在于这些宠物主人的收入较低[15,71],也与他们自身的健康和饮食习惯有关[71]。肥胖犬的主人也通常有肥胖问题方面的困扰[8,13,17,71,73]。他们可能很少运动，也很少注意饮食健康。吃低热量饮食的主人养的犬往往不肥胖[17]。但是，猫的超重程度与其主人的体重指数之间没有相关性[60,73]。研究还发现，中、老年人比40岁以下成年人养的犬患肥胖症的风险更高[8,13,15]。这是由于老年人本身缺乏运动[8]，但犬主人的年龄并未影响犬的运动量[15]。此外，老年人往往拥有年龄较大的犬。同样，猫主人的年龄并不是导致猫肥胖的真正危险因素，因为猫主人的年龄与猫的年龄并没有明显的相关性[28]。另外，人们也许会认为孩子们可能会给猫太多的食物或把宠物食盆放得太满；然而，有报道称与孩子一起生活可以降低猫患肥胖症的风险[28]。此外，只有1只犬[11]或只有1~2只猫[22,24]的家庭有较高的患肥胖症风险，而同时养犬也养猫的家庭，猫患肥胖症的风险会降低[26]。

宠物主人低估了宠物的体况[8,13,15,26,28,29,31,60,65,66]是导致喂食过多的另一个原因，也被证实是导致猫肥胖症的主要因素[26,28,29,31]。例如，那些本来体重正常的猫，却被主人认为体重过轻，就可能会被喂食更多食物[28]，如果主人不认为他们的宠物超重，那么就没有理由去减少宠物的能量摄入[31]。同样，尽管宠物医生告知他们的犬超重，但主人可能会不愿意接受这个诊断结果，或者认为这不是一个严重的问题[65]。

1.4 结论

宠物肥胖症涉及许多风险因素，包括动物自身因素和人为因素，这些因素又往往导致另外一些棘手的疾病。在每一次兽医临床健康检查中，应对每只宠物进行全面的营养评估来确定风险因素，从而防止体重增加或实施减肥计划。针对动物自身因素，可以计算宠物每日个性化的能量需要。然而，最重要的是对宠物主人的观念进行教育，并消除人为因素（如日粮、运动量、主人态度和家庭特征），这些因素才是导致肥胖症流行的最主要原因。

参考文献

1. German AJ. The growing problem of obesity in dogs and cats. *The Journal of Nutrition* 2006;136(7 Suppl):1940S–1946S.
2. Laflamme DP. Understanding and managing obesity in dogs and cats. *Veterinary Clinics of North America: Small Animal Practice* 2006;36(6):1283–1295, vii.
3. Laflamme DP. Companion animals symposium: Obesity in dogs and cats: What is wrong with being fat? *Journal of Animal Science* 2012;90(5):1653–1662.

4. Burkholder WJ. Use of body condition scores in clinical assessment of the provision of optimal nutrition. *Journal of the American Veterinary Medical Association* 2000;217(5):650–654.
5. Laflamme D. Development and validation of a body condition score system for dogs. *Canine Practice* 1997;22(4):10–15.
6. Laflamme D. Development and validation of a body condition score system for cats: A clinical tool. *Feline Practice* 1997;25(5–6):13–18.
7. Krook L, Larsson S, Rooney JR. The interrelationship of diabetes mellitus, obesity, and pyometra in the dog. *American Journal of Veterinary Research* 1960;21:120–127.
8. Mason E. Obesity in pet dogs. *The Veterinary Record* 1970;86(21): 612–616.
9. Edney AT, Smith PM. Study of obesity in dogs visiting veterinary practices in the United Kingdom. *The Veterinary Record* 1986;118(14):391–396.
10. Lund EM, Armstrong PJ, Kirk CA, Kolar LM, Klausner JS. Health status and population characteristics of dogs and cats examined at private veterinary practices in the United States. *Journal of the American Veterinary Medical Association* 1999;214(9):1336–1341.
11. Robertson ID. The association of exercise, diet and other factors with owner-perceived obesity in privately owned dogs from metropolitan Perth, WA. *Preventive Veterinary Medicine* 2003;58(1–2):75–83.
12. McGreevy PD, Thomson PC, Pride C, Fawcett A, Grassi T, Jones B. Prevalence of obesity in dogs examined by Australian veterinary practices and the risk factors involved. *The Veterinary Record* 2005;156(22):695–702.
13. Colliard L, Ancel J, Benet JJ, Paragon BM, Blanchard G. Risk factors for obesity in dogs in France. *Journal of Nutrition* 2006;136(7):1951S–1954S.
14. Lund EM, Armstrong PJ, Kirk CA, Klausner JS. Prevalence and risk factors for obesity in adult dogs from private US veterinary practices. *International Journal of Applied Research in Veterinary Medicine* 2006;4(2):177–186.
15. Courcier EA, Thomson RM, Mellor DJ, Yam PS. An epidemiological study of environmental factors associated with canine obesity. *Journal of Small Animal Practice* 2010;51(7):362–367.
16. Sallander M, Hagberg M, Hedhammar A, Rundgren M, Lindberg JE. Energy-intake and activity risk factors for owner-perceived obesity in a defined population of Swedish dogs. *Preventive Veterinary Medicine* 2010;96(1–2):132–141.
17. Heuberger R, Wakshlag J. The relationship of feeding patterns and obesity in dogs. *Journal of Animal Physiology and Animal Nutrition* 2011;95(1):98–105.
18. Mao J, Xia Z, Chen J, Yu J. Prevalence and risk factors for canine obesity surveyed in veterinary practices in Beijing, China. *Preventive Veterinary Medicine* 2013;112(3–4):438–442.
19. Corbee RJ. Obesity in show dogs. *Journal of Animal Physiology and Animal Nutrition* 2013;97(5):904–910.
20. Such ZR, German AJ. Best in show but not best shape: A photographic assessment of show dog body condition. *The Veterinary Record* 2015;177(5):125.

21. Sloth C. Practical management of obesity in dogs and cats. *Journal of Small Animal Practice* 1992;33(4):178–182.
22. Scarlett JM, Donoghue S, Saidla J, Wills J. Overweight cats: Prevalence and risk-factors. *International Journal of Obesity* 1994;18:S22–SS8.
23. Donoghue S, Scarlett JM. Diet and feline obesity. *Journal of Nutrition* 1998;128(12 Suppl):2776S–2778S.
24. Robertson ID. The influence of diet and other factors on owner-perceived obesity in privately owned cats from metropolitan Perth, Western Australia. *Preventive Veterinary Medicine* 1999;40(2):75–85.
25. Russell K, Sabin R, Holt S, Bradley R, Harper EJ. Influence of feeding regimen on body condition in the cat. *The Journal of Small Animal Practice* 2000;41(1):12–17.
26. Allan FJ, Pfeiffer DU, Jones BR, Esslemont DHB, Wiseman MS. A cross-sectional study of risk factors for obesity in cats in New Zealand. *Preventive Veterinary Medicine* 2000;46(3):183–196.
27. Lund EM, Armstrong PJ, Kirk CA, Klausner JS. Prevalence and risk factors for obesity in adult cats from private US veterinary practices. *International Journal of Applied Research in Veterinary Medicine* 2005;3(2):88–96.
28. Colliard L, Paragon BM, Lemuet B, Bénet JJ, Blanchard G. Prevalence and risk factors of obesity in an urban population of healthy cats. *Journal of Feline Medicine and Surgery* 2009;11(2):135–140.
29. Courcier EA, O'Higgins R, Mellor DJ, Yam PS. Prevalence and risk factors for feline obesity in a first opinion practice in Glasgow, Scotland. *Journal of Feline Medicine and Surgery* 2010;12(10):746–753.
30. Courcier EA, Mellor DJ, Pendlebury E, Evans C, Yam PS. An investigation into the epidemiology of feline obesity in Great Britain: Results of a cross-sectional study of 47 companion animal practises. *The Veterinary Record* 2012;171(22):560.
31. Cave NJ, Allan FJ, Schokkenbroek SL, Metekohy CA, Pfeiffer DU. A cross-sectional study to compare changes in the prevalence and risk factors for feline obesity between 1993 and 2007 in New Zealand. *Preventive Veterinary Medicine* 2012;107(1–2):121–533.
32. Corbee RJ. Obesity in show cats. *Journal of Animal Physiology and Animal Nutrition* 2014;98(6):1075–1080.
33. Bland IM, Guthrie-Jones A, Taylor RD, Hill J. Dog obesity: Veterinary practices' and owners' opinions on cause and management. *Preventive Veterinary Medicine* 2010;94(3–4):310–315.
34. Jeusette I, Greco D, Aquino F, Detilleux J, Peterson M, Romano V et al. Effect of breed on body composition and comparison between various methods to estimate body composition in dogs. *Research in Veterinary Science* 2010;88(2):227–232.
35. Arch JR, Hislop D, Wang SJ, Speakman JR. Some mathematical and technical issues in the measurement and interpretation of open-circuit indirect calorimetry in small animals. *International Journal of Obesity* 2006;30(9):1322–1331.

36. Kienzle E, Moik K. A pilot study of the body weight of pure-bred client-owned adult cats. *British Journal of Nutrition* 2011;106(1 Suppl):S113–S115.
37. Harper EJ. Changing perspectives on aging and energy requirements: Aging and energy intakes in humans, dogs and cats. *Journal of Nutrition* 1998;128(12 Suppl):2623S–2626S.
38. Harper EJ. Changing perspectives on aging and energy requirements: Aging, body weight and body composition in humans, dogs and cats. *Journal of Nutrition* 1998;128(12 Suppl):2627S–2631S.
39. Bermingham EN, Thomas DG, Cave NJ, Morris PJ, Butterwick RF, German AJ. Energy requirements of adult dogs: A meta-analysis. *PLOS ONE* 2014;9(10):e109681.
40. Bermingham EN, Weidgraaf K, Hekman M, Roy NC, Tavendale MH, Thomas DG. Seasonal and age effects on energy requirements in domestic short-hair cats (*Felis catus*) in a temperate environment. *Journal of Animal Physiology and Animal Nutrition* 2013;97(3):522–530.
41. Kealy RD, Lawler DF, Ballam JM, Mantz SL, Biery DN, Greeley EH et al. Effects of diet restriction on life span and age-related changes in dogs. *Journal of the American Veterinary Medical Association* 2002;220(9):1315–1320.
42. Lauten SD, Cox NR, Brawner WR, Baker HJ. Use of dual energy x-ray absorptiometry for noninvasive body composition measurements in clinically normal dogs. *American Journal of Veterinary Research* 2001;62(8):1295–1301.
43. Bermingham EN, Thomas DG, Morris PJ, Hawthorne AJ. Energy requirements of adult cats. *British Journal of Nutrition* 2010;103(8):1083–1093.
44. Spain CV, Scarlett JM, Houpt KA. Long-term risks and benefits of early- age gonadectomy in dogs. *Journal of the American Veterinary Medicine Association* 2004;224(3):380–387.
45. Spain CV, Scarlett JM, Houpt KA. Long-term risks and benefits of early- age gonadectomy in cats. *Journal of the American Veterinary Medical Association* 2004;224(3):372–379.
46. Houpt KA, Coren B, Hintz HF, Hilderbrant JE. Effect of sex and reproductive status on sucrose preference, food intake, and body weight of dogs. *Journal of the American Veterinary Medical Association* 1979;174(10):1083–1085.
47. Fettman MJ, Stanton CA, Banks LL, Hamar DW, Johnson DE, Hegstad RL et al. Effects of neutering on bodyweight, metabolic rate and glucose tolerance of domestic cats. *Research in Veterinary Science* 1997;62(2):131–136.
48. Harper EJ, Stack DM, Watson TDG, Moxham G. Effects of feeding regimens on bodyweight, composition and condition score in cats following ovariohysterectomy. *Journal of Small Animal Practice* 2001;42(9):433–438.
49. Belsito KR, Vester BM, Keel T, Graves TK, Swanson KS. Impact of ovariohysterectomy and food intake on body composition, physical activity, and adipose gene expression in cats. *Journal of Animal Science* 2009;87(2):594–602.

50. Flynn MF, Hardie EM, Armstrong PJ. Effect of ovariohysterectomy on maintenance energy requirement in cats. *Journal of the American Veterinary Medical Association* 1996;209(9):1572–1581.
51. Jeusette I, Detilleux J, Cuvelier C, Istasse L, Diez M. Ad libitum feeding following ovariectomy in female Beagle dogs: Effect on maintenance energy requirement and on blood metabolites. *Journal of Animal Physiology and Animal Nutrition* 2004;88(3–4):117–121.
52. Root MV, Johnston SD, Olson PN. Effect of prepuberal and postpuberal gonadectomy on heat production measured by indirect calorimetry in male and female domestic cats. *American Journal of Veterinary Research* 1996;57(3):371–374.
53. Martin L, Siliart B, Dumon H, Backus R, Biourge V, Nguyen P. Leptin, body fat content and energy expenditure in intact and gonadectomized adult cats: A preliminary study. *Journal of Animal Physiology and Animal Nutrition* 2001;85(7–8):195–199.
54. Hart BL, Barrett RE. Effects of castration on fighting, roaming, and urine spraying in adult male cats. *Journal of the American Veterinary Medical Association* 1973;163(3):290–292.
55. Hopkins SG, Schubert TA, Hart BL. Castration of adult male dogs: Effects on roaming, aggression, urine marking, and mounting. *Journal of the American Veterinary Medical Association* 1976;168(12):1108–1110.
56. Romsos DR, Belo PS, Bennink MR, Bergen WG, Leveille GA. Effects of dietary carbohydrate, fat and protein on growth, body composition and blood metabolite levels in the dog. *Journal of Nutrition* 1976;106(10):1452–1464.
57. Backus RC, Cave NJ, Keisler DH. Gonadectomy and high dietary fat but not high dietary carbohydrate induce gains in body weight and fat of domestic cats. *British Journal of Nutrition* 2007;98(3):641–650.
58. Nguyen PG, Dumon HJ, Siliart BS, Martin LJ, Sergheraert R, Biourge VC. Effects of dietary fat and energy on body weight and composition after gonadectomy in cats. *American Journal of Veterinary Research* 2004;65(12):1708–1713.
59. Bland IM, Guthrie-Jones A, Taylor RD, Hill J. Dog obesity: Owner attitudes and behaviour. *Preventive Veterinary Medicine* 2009;92(4): 333–340.
60. Kienzle E, Bergler R. Human-animal relationship of owners of normal and overweight cats. *Journal of Nutrition* 2006;136(7 Suppl):1947S–1950S.
61. LeBlanc J, Diamond P. Effect of meal size and frequency on postprandial thermogenesis in dogs. *American Journal of Physiology* 1986;250(2 Pt 1): E144–E147.
62. Deng P, Iwazaki E, Suchy SA, Pallotto MR, Swanson KS. Effects of feeding frequency and dietary water content on voluntary physical activity in healthy adult cats. *Journal of Animal Science* 2014;92(3): 1271–1277.
63. Slater MR, Robinson LE, Zoran DL, Wallace KA, Scarlett JM. Diet and exercise patterns in pet dogs. *Journal of the American Veterinary Medical Association* 1995;207(2):186–190.
64. Courcier EA, Mellor DJ, Thomson RM, Yam PS. A cross sectional study of the prevalence and risk factors for owner misperception of canine body shape in first opinion practice in Glasgow. *Preventive Veterinary Medicine* 2011;102(1):66–74.

65. White GA, Hobson-West P, Cobb K, Craigon J, Hammond R, Millar KM. Canine obesity: Is there a difference between veterinarian and owner perception? *Journal of Small Animal Practice* 2011;52(12):622–626.
66. Eastland-Jones RC, German AJ, Holden SL, Biourge V, Pickavance LC. Owner misperception of canine body condition persists despite use of a body condition score chart. *Journal of Nutritional Science* 2014;3:e45.
67. German AJ, Holden SL, Mason SL, Bryner C, Bouldoires C, Morris PJ et al. Imprecision when using measuring cups to weigh out extruded dry kibbled food. *Journal of Animal Physiology and Animal Nutrition* 2011;95(3):368–373.
68. Larsson C, Junghans P, Tauson AH. Evaluation of the oral 13C-bicarbonate tracer technique for the estimation of CO_2 production and energy expenditure in dogs during rest and physical activity. *Isotopes in Environmental and Health Studies* 2010;46(4):432–443.
69. Degeling C, Burton L, McCormack GR. An investigation of the association between socio-demographic factors, dog-exercise requirements, and the amount of walking dogs receive. *Canadian Journal of Veterinary Research* 2012;76(3):235–240.
70. Chan CB, Spierenburg M, Ihle SL, Tudor-Locke C. Use of pedometers to measure physical activity in dogs. *Journal of the American Veterinary Medical Association* 2005;226(12):2010–2015.
71. Kienzle E, Bergler R, Mandernach A. A comparison of the feeding behavior and the human–animal relationship in owners of normal and obese dogs. *Journal of Nutrition* 1998;128(12 Suppl):2779S–2782S.
72. Suarez L, Peña C, Carretón E, Juste MC, Bautista-Castaño I, Montoya-Alonso JA. Preferences of owners of overweight dogs when buying commercial pet food. *Journal of Animal Physiology and Animal Nutrition* 2012;96(4):655–659.
73. Nijland ML, Stam F, Seidell JC. Overweight in dogs, but not in cats, is related to overweight in their owners. *Public Health Nutrition* 2010;13(1):102–106.
74. German AJ, Holden SL, Moxham GL, Holmes KL, Hackett RM, Rawlings JM. A simple, reliable tool for owners to assess the body condition of their dog or cat. *Journal of Nutrition* 2006;136(7 Suppl):2031S–2033S.
75. Sallander M, Hedhammar A, Rundgren M, Lindberg JE. Demographic data of a population of insured Swedish dogs measured in a questionnaire study. *Acta Veterinaria Scandinavica* 2001;42(1):71–80.
76. Sallander MH, Hedhammar A, Rundgren M, Lindberg JE. Repeatability and validity of a combined mail and telephone questionnaire on demographics, diet, exercise and health status in an insured-dog population. *Preventive Veterinary Medicine* 2001;50(1–2):35–51.
77. Gant PH, Holden SL, Biourge V, Morris PJ, German AJ. Can body composition in dogs be estimated from photographs? *Journal of Veterinary Internal Medicine* 2013;27:742.

2 肥胖症的病理生理学：代谢效应和炎症介质

2.1 能量消耗与能量摄入的平衡

一般情况下，肥胖症是由能量摄入和消耗之间的不平衡引起的。然而，导致能量消耗增加或能量消耗减少的因素是复杂多样的。那么，为什么肥胖症在人类和伴侣动物中都上升到了流行病的程度呢？在多项评估人类双胞胎与家庭成员关系的研究中发现，遗传因素约占受试人群肥胖症和体重指数（body mass index，BMI）差异的65%[1-6]。其余大部分体重指数的差异似乎是由个体差异引起的，而非共同的环境效应[1]。对犬而言，某些品种（见第1章）患肥胖症的概率比其他品种更高[7]。虽然遗传因素的确对个体的胖瘦程度有一定作用，但在过去的50年中，肥胖率的急剧上升并不能完全归因于遗传变化。自20世纪60年代初以来，美国成年人肥胖率增加了1倍以上（从13.4%上升到35.7%）[8,9]。大多数专家已经形成共识，随着现代社会的发展，充足的食物供应和久坐行为的增加，与"肥胖易感基因"相结合，完美创造了"胖子的世界"，且有一些理论可以进一步阐明这种关系。

2.1.1 肥胖症的进化理论

节俭基因假说（thrifty gene hypothesis）是最早解释现代肥胖症的理论之一[10]。这种观点认为在饥荒期间，自然选择对瘦弱的个体施加了压力。在过去，那些新陈代谢慢、脂肪储存能力强的个体更容易在恶劣的环境中生存下来。此外，在饥荒时期，较高的体脂水平可以更好地维持生育能力。然而，该理论的一个潜在缺陷是，在人类进化过程中，自然选择哪怕稍微"偏向"肥胖的遗传突变，就会导致大约99%的现代人受肥胖症的影响[11]。

另一种理论认为，瘦弱的个体在进化上具有优势，因为他们能够更快地避开捕食者。随着社会的发展，食物供应趋于稳定，捕食活动减少，使更多拥有不同体型的个体得以生存繁衍。这种选择的"放松"导致了遗传漂变（genetic drift），从而产生了肥胖表型，或"漂变基因（drifty genes）"[11]。

众所周知，某些人类种族和少数民族的肥胖率高于其他种族。解释该现象的最新假说认为，这与相应族群体内拥有的棕色（褐色）脂肪组织（brown adipose tissue,BAT)和解偶联蛋白 1(uncoupling protein 1,UCP1)基因有关。棕色脂肪组织和解偶联蛋白 1 都可以调节能量以热量的形态释放。婴儿和儿童的棕色脂肪组织含量很高,这对于其存活和产热很重要。此外,生活在寒冷气候中的人和动物往往含有更多的棕色脂肪组织(BAT),且解偶联蛋白 1(UCP1)的表达量更高。另外,能量以热的形式释放也会提高代谢率。在温暖气候中进化的人类可能新陈代谢较慢,这与体内棕色脂肪组织含量和解偶联蛋白 1 表达量较低有关。肥胖症进化的这种"产热(thermogenic)"理论可以解释为什么在美洲印第安人、黑色人种和西班牙裔人群中发现有较高的肥胖率[12,13]。

当按照现代新型饮食或活动模式时,体重具有增加的遗传倾向,大多数人类肥胖症进化理论都集中于此。近年来,犬猫肥胖症的攀升也可能是这些环境模式变化的结果。大多数伴侣动物都可以轻松获得极其美味和高能量的食物。此外,与犬猫的祖先相比,现在的犬猫更多的是待在室内,很少有运动锻炼的机会。一种普遍的理论认为,猫已经进化成为严格的食肉动物,采食高碳水化合物含量的商业猫粮导致了猫肥胖症大范围流行。然而有证据表明,较低的碳水化合物日粮(35 g/1 000 kcal vs. 76 g/1 000 kcal)①可以更好地调节患有糖尿病的猫的血糖浓度[14],并且可以降低健康猫的餐后胰岛素和血糖浓度[15,16],但是目前尚无证据表明日粮中碳水化合物(CHO)的含量可以直接影响肥胖症发生的风险。与之相反,有些研究发现,与喂食高脂肪(63 g/1 000 kcal,54 g/1 000 kcal CHO)或高蛋白质(124 g/1 000 kcal,62 g/1 000 kcal CHO)日粮的猫相比,喂食高碳水化合物(126~137 g/1 000 kcal,58~80 g/1 000 kcal 蛋白质,29~32 g/1 000 kcal 脂肪)日粮的猫,会增加脂肪沉积并且消耗更多热量[15,17]。目前,尚不清楚长期喂食高碳水化合物日粮是否会导致猫患糖尿病。虽然需要更多的研究来阐明碳水化合物在肥胖症发展过程中的作用,但有些环境因素,如生活在室内和随意喂食,确实发现在肥胖猫上发生的频率比瘦猫更高;然而,并不是所有报道都发现了与此一致的现象[18-20]。例如,一项以干粮为主的饮食对肥胖症是否有影响的研究发现,两者存在显著相关性,而另一项研究则没有发现相关性[19,21](见第 1 章)。也有几项研究未能观察到喂食方式或户外活动与肥胖症之间的相关性[22-24]。

关于自然进化和现代社会变化的理论并不是解释肥胖症增多的唯一假说。目前正在研究的,用以解释肥胖症急剧增加的其他理论因素包括传染源(例如病毒和细菌)、肠道菌群改变(请参见第 2.3 节)、环境污染物对内分泌的破坏,以及社会心理压力[25]。

此外,食欲和采食通过中枢调控、外周调控和享乐调节的方式也必须加以考虑。

---

① 1 cal=4.184 J;1 000 kcal=1 Mcal

2.1.2 食欲和采食调节

2.1.2.1 食欲的中枢调节

虽然中枢神经系统对食欲和采食的调节是一个复杂的过程,但主要的调控中心是下丘脑弓状核。该区域的神经元表达阿黑皮素原(proopiomelanocortin,POMC)和可卡因-苯丙胺调节转录物(Cocaine- and amphetamine-regulated transcript,CART)。这些信号分子可降低食欲,增加能量消耗,通常被描述为抑食神经肽(anorexigenic)。也有神经元表达刺鼠相关蛋白(agouti-related protein,AgRP)和神经肽Y(neuropeptide Y,NPY),可增加食欲,同时减少能量消耗,这些神经元被称为促食神经肽(orexigenic)(图2.1)。来自胃和脂肪组织等外周组织的激素信号通过刺激下丘脑,激发或抑制神经元影响食欲。弓状核内的神经元还与下丘脑其他区域的"抑食神经肽"和"促食神经肽"进行信息传递,以调节采食。

2.1.2.2 食欲的外周调节

为了确定身体何时需要更多或更少的食物,大脑和下丘脑必须依靠其他器官和组织来传达瞬时性的和长期性的能量需要信号。例如,为防止胃部过度膨胀,胃中的牵张感受器(stretch receptors)可以向大脑发送信号来结束一次进餐;而脂肪组织会发送激素信号来"告诉"大脑何时增加或减少脂肪存储。其中研究最透彻的脂肪激素是瘦素(leptin)。当一个人获得多余脂肪时,脂肪细胞就会释放更多的瘦素。瘦素进入下丘脑,刺激厌食神经元减少采食,增加能量消耗。另外,如果一个人正在减肥,脂肪储存减少,瘦素浓度较低,以及阿黑皮素原(POMC)和可卡因-

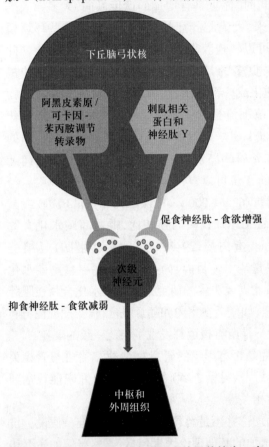

图2.1 食欲调控中枢存在于下丘脑弓状核内。表达阿黑皮素原(POMC)和可卡因-苯丙胺调节转录物(CART)的神经元,可降低食欲,增加能量消耗,称为抑食神经肽。表达刺鼠相关蛋白(AgRP)和神经肽Y(NPY)的神经元,可增加食欲,同时减少能量消耗,称为促食神经肽

苯丙胺调节转录物（CART）的释放减少，因此会增加食欲，减缓代谢速度。因此，瘦素的作用是维持体脂平衡，也被称为调节脂肪含量的"恒脂器"（adipostat），与恒温器原理有些相似[27]。瘦素除了发出长期能量储存的信号外，还可以发出能让能量摄入急剧减少的信号。例如，一项对健康男性禁食 2~3 d 的研究表明，血清瘦素浓度会显著降低[28]。尽管瘦素试图通过调控食欲来防止体重过度增加，但肥胖人群的瘦素浓度往往更高，并且可能对瘦素的作用产生抵抗。这类似于胰岛素抵抗，虽然血液循环中的瘦素浓度水平很高，但是瘦素抵抗作用仍会导致其失效[29]。一些间接证据表明，瘦素抵抗也可能发生在犬和猫身上。许多研究表明，随着脂肪含量的增加，犬和猫体内循环的瘦素水平升高，但尚无研究表明瘦素与自由采食有关[30-36]。

除了脂肪组织外，肠道向大脑发送采食信号也至关重要。如前所述，胃和肠道中的牵张感受器通过进食进而发出膨胀信号，并通过迷走神经传递信号刺激食欲中枢。在含有能量的营养物质出现在肠道时，肠道会释放许多肠肽，这些营养物质主要是日粮中小分子蛋白质和氨基酸。除了从胃中释放的胃饥饿素能刺激饥饿和促进觅食行为外，其他所有已知的肠肽都会抑制食欲。胆囊收缩素（cholecystokinin，CCK）、酪酪肽（peptide YY，PYY）和胰高血糖素样肽-1（glucagon-like peptide-1，GLP-1）是研究最深入的可调节食欲的 3 种肠肽，它们通过刺激迷走神经和脑干，间接刺激下丘脑区域[37]。

2.1.2.3 食欲的快感（享乐）调控

虽然下丘脑和大脑的其他区域是依赖生理信号来调控食欲的，但目前对人类的研究表明，奖赏、注意力、情绪和其他感知调控系统相关的大脑高级中枢，在调节进食过程中也是非常重要的[26]。这些高级中枢与下丘脑直接联系，常常可以超越生理饱腹感信号。

吃东西可以使人愉悦并刺激大脑的"奖赏中心"。然而，在肥胖状态下，食物奖赏的作用会发生改变。一种理论认为，肥胖者大脑纹状体多巴胺 2 受体（dopamine 2 receptor）的有效性较低。因此，需要更"高回报"的食物（高脂肪或高热量）来获得满足感。然而，一些证据表明，摄入高脂肪或高热量食物本身就是导致多巴胺 2 受体水平降低的原因。这种对食物奖赏的低反应性（hyporesponsiveness）理论通常等同于食物成瘾，类似于药物成瘾。将食物奖赏与肥胖联系起来的另一种理论认为，接触"高回报"食物会导致奖赏中心的过度反应，并导致个体渴望更多的食物。虽然这两种理论都证明了奖赏系统在食欲不振和肥胖中的作用，但还需要更多的对照研究[26]。

对人类的研究已经充分证明，情绪可以改变食欲。例如，焦虑和抑郁是肥胖症常见的合并症。与悲伤和恐惧相比，愤怒和喜悦会增加食欲，并导致非理性的食物选择和摄入[38]。虽然很难将这些发现完全外推到伴侣动物上，但我们知道，犬和猫有复杂的情感，它们的情感状态也可能在一定程度上影响能量摄入。

2.1.3 能量消耗和能量代谢调控机制

如本章开头所述,肥胖症是由能量摄入和能量消耗之间的不平衡引起的。维持能量需要(maintenance energy requirement,MER)是每天维持恒定体重所需的能量,由以下几部分组成[39]:

- 基础代谢
- 产热
- 活动量
- 合成代谢 - 生长发育、妊娠、哺乳

动物通过逐步氧化蛋白质、脂肪和碳水化合物的方式,逐渐释放出可供身体利用的能量,并产生二氧化碳、水和热量。这个过程称为分解代谢(Catabolism)。在体内,高能磷酸键以脂肪、蛋白质和碳水化合物形式存储,这种能量存储形成的过程称为合成代谢(Anabolism)[40]。

动物在热中性环境中,静卧和清醒状态下,绝食12 h后,维持所消耗的能量,被定义为基础代谢率(basal metabolic rate,BMR),基础代谢率占总能量消耗的很大部分,这是维持血液循环、呼吸和细胞代谢活动所需的能量。兽医营养学家经常使用与基础代谢率类似的另一个词——"静息能量需要(resting energy requirement,RER)"来描述。通常使用克莱伯定律(译者注:原文为Kleiber interspecies formula,直译为克莱伯物种间公式)来估算不同物种的基础代谢率(BMR)和静息能量需要(RER):体重$_{(kg)}^{0.75} \times 70$[41]。无脂肪组织(如器官和肌肉等无脂肪组织)含量是BMR和RER的最佳预测指标,因为脂肪组织的代谢仅占基础能量代谢中很小的一部分[39]。因此,上述公式最适用于理想体重的动物,而非超重动物[42]。

食物在消化和处理过程中消耗能量称为食物热效应(thermic effect of food,TEF),其可影响基础代谢,并在外界环境变化的情况下用于保持稳定的核心体温。基础代谢和产热受多种因素的影响。甲状腺及甲状腺激素在代谢活跃的组织,如肝脏、棕色脂肪组织、骨骼肌和心脏等的能量消耗中起重要作用。甲状腺激素还可通过降低热力学效率来提高能量的产热作用。这导致机体产热量增加并且伴随体温升高。当环境温度降到动物的热中性区(thermoneutral zone)以下时,大多数哺乳动物都会启动诸如发抖、血管收缩和竖毛等机制用以蓄热(heat-save)和产生热量。对于长期产热,哺乳动物也会利用棕色脂肪组织(棕色脂肪细胞中含有一种线粒体蛋白)产热。棕色脂肪组织的产热受到下丘脑的调节,去甲肾上腺素是主要的神经递质,甲状腺激素起辅助作用[43]。当食用含蛋白质、脂肪和碳水化合物的食物时,摄入的能量中大约10%将流向食物热效应(TEF)[39]。基础代谢率(MER)中变化最大的部分是身体活动期间的能量消耗。犬的基础代谢率（MER）变化范围很大（94~250 kcal × $BW_{kg}^{0.75}$），其中大部分差异是由于日常活动和运动上的差异所致[44]。

2.2 肥胖症的炎症效应

2.2.1 脂肪组织的内分泌功能

脂肪组织细胞释放多种内分泌、旁分泌和自分泌信号，这些信号称为脂肪因子（adipokine）。脂肪因子一词通常用于描述脂肪组织释放的任何蛋白质，无论是由脂肪细胞还是非脂肪细胞释放的[45]。尽管已发现许多种脂肪因子，但大多数脂肪因子的功能和生理相关性尚不明确。对一些脂肪因子进行深入研究发现，这些脂肪因子似乎对胰岛素敏感性有正向或负向的调节作用。大多数脂肪因子复杂的代谢作用尚不完全清楚，目前研究最深入的脂肪因子包括瘦素、脂联素（adiponectin）和肿瘤坏死因子α（tumor necrosis factor alpha，TNF-α）。

瘦素是第1个发现的脂肪因子，其主要功能是通过调节食欲和增加能量代谢来调控体脂含量（参见食欲的外周调控部分）。虽然瘦素的主要生理作用是调节体内脂肪的储存，但也会影响免疫、心血管和生殖系统[46,47]。瘦素还可以增强胰岛素信号传导，从而改善细胞内葡萄糖的利用并减少周围组织中脂质的累积[48]。不能产生瘦素的基因突变小鼠会发展成肥胖症、胰岛素抵抗和糖尿病，这种小鼠被称为 *ob/ob* 小鼠[49,50]。另外，瘦素具有促炎和促有丝分裂（致癌）的特性，因为瘦素可以刺激机体产生各种促炎细胞因子（见第2.2.2节对炎症的讨论）[51]。

脂联素是血液循环中分泌最多的脂肪因子[52]。尽管脂肪细胞负责分泌脂联素，但随着脂肪含量的增加，脂联素水平反而会下降[53]，导致这种异常变化的机理尚不明确。有研究推测，其他脂肪因子（如TNF-α）水平的升高可能会抑制脂联素的表达。脂联素影响多种代谢途径，其最主要的作用也许是作为胰岛素增敏剂。脂联素与胰岛素敏感性密切相关，与体脂含量无关[54-56]。对人类的前瞻性和纵向研究表明，较低水平的脂联素与胰岛素抵抗和糖尿病的未来发展密切相关[57-59]。健康成年人患2型糖尿病的风险较低，这与较高水平的脂联素密切相关[60]。许多临床和流行病学研究都将低水平的脂联素与肥胖症、胰岛素抵抗、2型糖尿病、高血压、心血管疾病和肝病等慢性炎症状态联系在一起[61,62]。

在伴侣动物中脂联素的研究得到了不同的结果。虽然大多数研究表明猫体内脂肪与脂联素呈负相关[34,63,65-67]，但其他研究表明其相关性较差[30,68,69]。检测技术和研究试验设计的差别可能是造成这些差异的原因。此外，对猫进行的大多数研究中检测都是以总脂联素（total adiponectin）的形式，而不是更活跃的高分子量形式,这可能会导致发现相关性的概率降低。迄今为止,仅有2项研究评估了脂联素的多聚体形式对猫体脂含量的影响,结果都发现脂联素的高分子量组分与体脂百分比的关系比总脂联素更紧密[30,68]。在犬中,脂联素的研究也发现了类似的结果,一些研究表明脂联素与体脂呈负相关[35,71,72,74],而另一些研究则没有发现相关关系[36,75-77]。

TNF-α是由多种细胞表达的炎性细胞因子,包括巨噬细胞、肥大细胞、神经元细胞、成纤维细胞和脂肪细胞。TNF-α、肥胖症和胰岛素抵抗之间的关系尚不明确。由于分化和未分化的脂肪细胞都可以分泌TNF-α,因此认为肥胖者中TNF-α水平的升高主要是由于脂肪细胞的分泌;然而,脂肪组织间质血管部分内的细胞,包括巨噬细胞,产生的TNF-α明显多于脂肪细胞[79-81]。人类的大多数脂肪TNF-α都会发挥局部旁分泌和自分泌作用[82,83]。对于小鼠和犬,更多的TNF-α被释放到体循环中[75,84]。来自脂肪组织的TNF-α在猫体内的循环模式尚不清楚,但其脂肪中mRNA表达量会随着肥胖而增加[86]。无论哪个物种,脂肪TNF-α的主要作用是诱导局部胰岛素抵抗。某些基因负责胰岛素介导的葡萄糖摄取进入细胞,而TNF-α可下调这些基因[87,89]。除了抑制葡萄糖进入细胞外,TNF-α还可以减少脂肪细胞对游离脂肪酸(free fatty acid,FFA)的摄取,并促进脂肪分解和游离脂肪酸释放进入循环系统[80,82,90]。这导致游离脂肪酸(FFA)水平在血液循环中增加,并且对外周组织中的胰岛素敏感性产生负面影响。除了直接影响脂肪组织的胰岛素敏感性外,TNF-α还可以改变参与葡萄糖代谢的其他脂肪因子的分泌。尤其是TNF-α与脂联素呈负相关,并可能改变其基因表达[80,82,91,92]。总之,脂肪组织分泌的TNF-α在局部和全身水平的糖脂代谢中起着重要作用,是肥胖相关炎症的关键组成部分。

2.2.2 肥胖症和炎症

众所周知,肥胖者往往生活在一种慢性、低度炎症状态中,这种炎症状态与过多的脂肪组织直接相关。这种炎性状态被认为加重了人类癌症和其他慢性疾病的发展。据估计,人类众多癌症种类中,有20%~35%由肥胖引起[93,94]。虽然肥胖症和癌症发展背后的机制具有肿瘤特异性,但许多癌症类型可能继发于炎症状态[93]。产生这种炎症状态的2种方式可以简化为脂肪细胞自身水平的局部炎症和脂肪组织炎症介质的全身释放。

2.2.2.1 局部炎症

脂肪细胞由多能干细胞(multipotent stromal stem cells,MSSCs)发育而来。脂肪的增加必然来自2种可能,一是含有更多甘油三酯的脂肪细胞的增大,二是脂肪细胞数量的增加。除非极端情况,否则成年人脂肪细胞数量相对稳定。然而,儿童脂肪细胞更容易增生,进而增加了他们一生中脂肪细胞的数量[95,96]。

众所周知,肥胖的一个副作用是炎症。一般认为脂肪组织内缺氧是诱发炎症的主要因素。针对肥胖的人和小鼠的研究已表明,缺氧可以减少脂肪组织的血流量。另外,氧气只能扩散通过约 120 μm 的组织,而在脂肪细胞中则可高达 150 μm[97]。肥胖小鼠的血管生成减少,血管收缩增加[97]。慢性炎症状态导致产生过量活性氧(reactive oxygen species,ROS)和全身性氧化应激。高水平的氧化应激与人类肥胖症、血管异常、动脉粥样硬化以及心血管疾病风险升高密切相关[98]。

2.2.2.2 全身炎症

如前所述,从脂肪组织释放的 2 种激素在全身炎症中起关键作用。抗炎激素脂联素的浓度随着脂肪含量的升高而降低,而促炎激素瘦素的浓度则升高。瘦素通过先天免疫系统刺激炎性细胞因子如 IL-6、IL-12、IL-1 和 TNF-α 的产生。瘦素还将刺激一氧化氮、环氧合酶 2(cyclooxygenase 2,COX2)和活性氧的产生[51]。除了改变脂肪因子的表达,肥胖还可以增强肾素-血管紧张素系统的活性,从而增加细胞内 ROS、线粒体功能障碍和 DNA 损伤[99]。

2.3 微生物在肥胖症中的作用

2.3.1 肠道微生物菌群与肥胖症

胃肠道微生物菌群对宿主健康有深远影响,并且肠道微生物区系与肥胖症之间已经建立起密切的联系[100,101]。大多数有关肥胖肠道菌群的研究都是在人类和啮齿动物上进行的。一般来说,肥胖者肠道菌群的特点是缺乏多样性[102-106]。在类群层次上的变化,例如同一物种的肥胖个体与瘦个体的肠道或粪便中厚壁菌门(*Firmicutes* phylum)成员的丰度,在不同的研究中是不同的。在为数不多已发表的关于犬的研究中发现,肥胖犬与瘦犬的特征缺乏一致性[107,108]。不同研究之间的差异可能是由于不同研究群体之间的差异,例如单独被宠物主人养的犬与群居的犬的差异,也可能是生活环境对肠道微生物产生的影响[109]。此外,这种变化也可以解释为在评估复杂肠道菌群(通常是粪便菌群)的方法上缺乏共识[110]。将肠道微生物用于疾病管理是一项挑战,由于广泛的细菌分类群之间存在"保守"功能序列,肠道菌群的成员隶属关系并不能直接反映其功能。肠道微生物可能影响肥胖的机制包括:改变日粮消化和能量吸收,增强脂肪合成和脂肪储存,以及食欲调节[100,105,111-114]。但这些机制尚未在犬或猫上进行研究验证。

2.3.2 肠道微生物菌群在肥胖症治疗中的应用

2.3.2.1 饮食策略

日粮对犬肠道菌群[115]和其他物种的肠道菌群均有影响。这并不奇怪,因为微生物可以高效快速适应底物的变化。一些抗肥胖日粮的干预措施(例如增加纤维、改变脂肪或增加多酚)与肠道菌群同步变化或预防菌群失调有关[116,117]。这些研究将日粮、肠道菌群与肥胖症联系起来,但是,肠道菌群与肥胖症之间的联系仍未完全阐明。此外,其他类似日粮干预措施的研究表明,日粮的影响与肠道菌群的变化无关[118,119]。目前,似乎还没有一种实用的方法能通过日粮对肠道微生物的动态调节来治疗犬肥胖症[120]。不过,关于日粮如何通过肠道微生物菌群调节肥胖症的研究仍在进行中。

2.3.2.2 非饮食策略

益生菌通常是指对宿主健康有益的、活的微生物。益生菌被认为可以调节宿主肠道菌群和改变代谢,从而可能成为治疗肥胖症的一种策略[121]。然而,在不同研究中,比较益生菌之间的有益效果是一项挑战,原因是益生菌的制备、菌株、菌株的数量和形态(粉状、咀嚼片状等)等方面均存在差异。犬益生菌补充剂中常用的益生菌包括乳酸杆菌(*Lactobacillus*)、双歧杆菌(*Bifidobacterium*)和肠球菌(*Enterococcu*)[121-123]。但是,这些常用的益生菌并未显示出有助于预防肥胖症的作用[124,125],甚至还可能导致体重增加[126]。通常认为益生菌对犬是安全的。然而,益生菌并非没有风险,在新生幼犬和老年患病犬中,益生菌与菌血症(bacteremia)和真菌血症(fungemia)有关[127,128]。

抗生素也能改变犬的肠道菌群[129],但由于对犬抗生素耐药性的担忧[130,131],这种非饮食策略可能被认为是不道德的。粪菌移植是一种将健康宿主肠道或粪便内容物中微生物转移到患病宿主的传统方法[132],该方法近期又开始流行[85,133,134]。迄今为止,粪菌移植对肥胖症的治疗并未看到良好前景,因为有较多证据表明粪菌移植将促进体重增加[73,78,100,119],而非体重减轻[64]。由于肠道菌群的建立始于出生前[70],而在成年期变化较小,因此,未来最有前途的方向,可能在于阐明宠物母亲和新生宠物的肠道菌群功能。

2.4 结论

虽然肥胖症可简单描述为能量摄入超过能量消耗的结果,但很显然,肥胖症与能量摄入和消耗之间的关系是复杂的,并受到各种因素的影响。患病宠物的基因组成和其所处环境,在过量脂肪组织的形成过程中都起着巨大的作用。例如,与青年威玛猎犬相比，中年拉布拉多犬由于其甲状腺功能减退，每千克体重可能要消耗更少的热量才能保持较瘦体型。我们现在也认识到，肥胖症对犬猫整个身体都有系统性影响，过多的脂肪组织产生的促炎状态会导致癌症和糖尿病等慢性疾病。通过更好地了解肥胖症的成因和后果，宠物医生能够更好地与宠物主人进行沟通，并制定有效的宠物减肥计划。

参考文献

1. Speakman JR, Westerterp KR. A mathematical model of **weight loss under total** starvation: Evidence against the thrifty-gene hypothesis. *Disease Models & Mechanisms* 2013;6(1):236–251.
2. Allison DB, Kaprio J, Korkeila M, Koskenvuo M, Neale MC, Hayakawa K. The heritability of body mass index among an international sample of monozygotic twins reared apart. *International Journal of Obesity and Related Metabolic Disorders: Journal of the International Association for the Study of Obesity* 1996;20(6):501–506.
3. Segal NL, Allison DB. Twins and virtual twins: Bases of relative body weight revisited. *International Journal of Obesity and Related Metabolic Disorders: Journal of the International Association for the Study of Obesity* 2002;26(4):437–441.
4. Luke A, Guo X, Adeyemo AA et al. Heritability of obesity-related traits among Nigerians, Jamaicans and US black people. *International Journal of Obesity and Related Metabolic Disorders: Journal of the International Association for the Study of Obesity* 2001;25(7):1034–1041.
5. Wu X, Cooper RS, Borecki I et al. A combined analysis of genomewide linkage scans for body mass index from the National Heart, Lung, and Blood Institute Family Blood Pressure Program. *American Journal of Human Genetics* 2002;70(5):1247–1256.
6. Zhu X, Cooper RS, Luke A et al. A genome-wide scan for obesity in African-Americans. *Diabetes* 2002;51(2):541–544.
7. Lund EM, Armstrong PJ, Kirk CA, Klausner JS. Prevalence and risk factors for obesity in adult dogs from private US veterinary practices. *International Journal of Applied Research in Veterinary Medicine* 2006;4:177–186.
8. Ogden CL, Carroll MD. *Prevalence of overweight, obesity, and extreme obesity among adults: United States, trends 1960–1962 through 2007– 2008. NCHS Health E-Stat.* Hyattsville, MD: National Center for Health Statistics; 2010. Available online: http://www.cdc.gov/NCHS/data/hestat/ obesity_adult_07_08/obesity_adult_07_08.pdf.
9. Flegal KM, Carroll MD, Kit BK, Ogden CL. Prevalence of obesity and trends in the distribution of body mass index among US adults, 1999– 2010. *Journal of the American Medical Association* 2012;307(5):491–497. Available online: http://jama.jamanetwork.com/article. aspx?articleid=1104933.

10. Neel J. Diabetes mellitus a "thrifty" genotype rendered detrimental by "progress"? *American Journal of Human Genetics* 1962;14:352–353.
11. Speakman JR. Thrifty genes for obesity, an attractive but flawed idea, and an alternative perspective: The 'drifty gene' hypothesis. *International Journal of Obesity* 2008;32(11):1611–1617.
12. Albuquerque D, Stice E, Rodriguez-Lopez R, Manco L, Nobrega C. Current review of genetics of human obesity: From molecular mechanisms to an evolutionary perspective. *Molecular Genetics and Genomics* 2015;290(4):1191–1221.
13. Sellayah D, Cagampang FR, Cox RD. On the evolutionary origins of obesity: A new hypothesis. *Endocrinology* 2014;155(5):1573–1588.
14. Bennett N, Greco DS, Peterson ME, Kirk C, Mathes M, Fettman MJ. Comparison of a low carbohydrate-low fiber diet and a moderate carbohydrate-high fiber diet in the management of feline diabetes mellitus. *Journal of Feline Medicine & Surgery* 2006;8(2):73–84.
15. Coradini M, Rand JS, Morton JM, Rawlings JM. Effects of two commercially available feline diets on glucose and insulin concentrations, insulin sensitivity and energetic efficiency of weight gain. *The British Journal of Nutrition* 2011;106(Suppl 1):S64–S77.
16. Farrow HA, Rand JS, Morton JM, O'Leary CA, Sunvold GD. Effect of dietary carbohydrate, fat, and protein on postprandial glycemia and energy intake in cats. *Journal of Veterinary Internal Medicine* 2013;27(5):1121–1135.
17. Gooding MA, Atkinson JL, Duncan IJH, Niel L, Shoveller AK. Dietary fat and carbohydrate have different effects on body weight, energy expenditure, glucose homeostasis and behaviour in adult cats fed to energy requirement. *Journal of Nutritional Science* 2015;4:e2.
18. Russell K, Sabin R, Holt S, Bradley R, Harper EJ. Influence of feeding regimen on body condition in the cat. *Journal of Small Animal Practice* 2000;41(1):12–17.
19. Harper EJ, Stack DM, Watson TD, Moxham G. Effects of feeding regimens on bodyweight, composition and condition score in cats following ovariohysterectomy. *The Journal of Small Animal Practice* 2001;42(9):433–438.
20. Kienzle E, Bergler R. Human-animal relationship of owners of normal and overweight cats. *The Journal of Nutrition* 2006;136(7 Suppl):1947S–1950S.
21. Rowe E, Browne W, Casey R, Gruffydd-Jones T, Murray J. Risk factors identified for owner-reported feline obesity at around one year of age: Dry diet and indoor lifestyle. *Preventive Veterinary Medicine* 2015;121(3–4):273–281.
22. Cave NJ, Allan FJ, Schokkenbroek SL, Metekohy CA, Pfeiffer DU. A cross-sectional study to compare changes in the prevalence and risk factors for feline obesity between 1993 and 2007 in New Zealand. *Preventive Veterinary Medicine* 2012;107(1–2):121–133.
23. Scarlett JM, Donoghue S, Saidla, Wills J. Overweight cats: Prevalence and risk factors. *International Journal Obesity Related Metabolic Disorders* 1994;S22–S28.

24. Allan FJ, Pfeiffer DU, Jones BR, Esslemont DHB, Wiseman MS. A cross-sectional study of risk factors for obesity in cats in New Zealand. *Preventive Veterinary Medicine* 2000;46(3):183–196.
25. Zinn A. Unconvential wisdom about the obesity epidemic. *The American Journal of the Medical Sciences* 2010;340(6):481–491.
26. Farr OM, Li CS, Mantzoros CS. Central nervous system regulation of eating: Insights from human brain imaging. *Metabolism: Clinical and Experimental* 2016;65(5):699–713.
27. Allison MB, Myers MG. 20 years of leptin: Connecting leptin signaling to biological function. *Journal of Endocrinology* 2014;223(1):T25–T35.
28. Chan JL, Heist K, DePaoli AM, Veldhuis JD, Mantzoros CS. The role of falling leptin levels in the neuroendocrine and metabolic adaptation to short-term starvation in healthy men. *The Journal of Clinical Investigation* 2003;111(9):1409–1421.
29. Zhou Y, Rui L. Leptin signaling and leptin resistance. *Frontiers of Medicine* 2013;7(2):207–222.
30. Witzel A, Kirk C, Kania S et al. Relationship of adiponectin and its multimers to metabolic indices in cats during weight change. *Domestic Animal Endocrinology* 2015;53:70–77.
31. Backus R, Havel P, Gingerich R, Rogers Q. Relationship between serum leptin immunoreactivity and body fat mass as estimated by use of a novel gas- phase Fourier transform infrared spectroscopy deuterium dilution method in cats. *American Journal of Veterinary Research* 2000;61(7):796–801.
32. Appleton D, Rand J, Sunvold G. Plasma leptin concentrations in cats: Reference range, effect of weight gain and relationship with adiposity as measured by dual energy x-ray absorptiometry. *Journal of Feline Medicine & Surgery* 2000;2(4):191–199.
33. Martin LJM, Siliart B, Dumon HJW, Nguyen P. Spontaneous hormonal variations in male cats following gonadectomy. *Journal of Feline Medicine & Surgery* 2006;8(5):309–314.
34. Hoenig M, Thomaseth K, Waldron M, Ferguson DC. Insulin sensitivity, fat distribution, and adipocytokine response to different diets in lean and obese cats before and after weight loss. *American Journal of Physiology: Regulatory, Integrative and Comparative Physiology* 2007;292(1):R227–R234.
35. Park HJ, Lee S, Oh JH, Seo KW, Song KH. Leptin, adiponectin and serotonin levels in lean and obese dogs. *BMC Veterinary Research* 2014;10:113–114.
36. Wakshlag JJ, Struble AM, Levine CB, Bushey JJ, Laflamme DP, Long GM. The effects of weight loss on adipokines and markers of inflammation in dogs. *British Journal of Nutrition* 2011;106(Suppl 1):S11–S14.
37. Ueno H, Nakazato M. Mechanistic relationship between the vagal afferent pathway, central nervous system and peripheral organs in appetite regulation. *Journal of Diabetes Investigation* 2016;7(6):812–818.
38. Macht M. Characteristics of eating in anger, fear, sadness and joy. *Appetite* 1999;33(1):129–139.

39. Butte N, Cabellero B. Energy needs: Assessment and requirements. In: Shils M, editor, *Modern Nutrition in Health and Disease*. 10th ed. New York: Lippincott Williams & Wilkins; 2006, p. 136–147.
40. Ganong W. *Review of Medical Physiology*. 22nd ed. New York: Lange Medical Books/McGraw-Hill; 2005.
41. Kleiber, M. *The Fire of Life*. John Wiley & Sons, New York; 1961.
42. Hill RC. Challenges in measuring energy expenditure in companion animals: A clinician's perspective. *The Journal of Nutrition* 2006;136(7 Suppl):1967S–1972S.
43. López M, Alvarez CV, Nogueiras R, Diéguez C. Energy balance regulation by thyroid hormones at central level. *Trends in Molecular Medicine* 2013;19(7):418–427.
44. National Research Council (NRC). Energy. In: *Nutrient Requirements of Dogs and Cats*. Washington, D.C.: The National Academies Press; 2006, chap. 3.
45. Fain JN, Tagele BM, Cheema P, Madan AK, Tichansky DS. Release of 12 adipokines by adipose tissue, nonfat cells, and fat cells from obese women. *Obesity* 2010;18(5):890–896.
46. Ren J. Leptin and hyperleptinemia – From friend to foe for cardiovascular function. *Journal of Endocrinology* 2004;181(1):1–10.
47. Munzberg H, Myers Jr. MG. Molecular and anatomical determinants of central leptin resistance. *Nature Neuroscience* 2005;8(5):566–570.
48. Park S, Hong SM, Sung SR, Jung HK. Long-term effects of central leptin and resistin on body weight, insulin resistance, and beta-cell function and mass by the modulation of hypothalamic leptin and insulin signaling. *Endocrinology* 2007. doi:10.1210/en.2007-0754.
49. Sell H, Dietze-Schroeder D, Eckel J. The adipocyte-myocyte axis in insulin resistance. *Trends in Endocrinology & Metabolism* 2006;17(10):416–422.
50. Halaas JL, Gajiwala KS, Maffei M et al. Weight-reducing effects of the plasma protein encoded by the obese gene. *Science* 1995;269(5223):543–546.
51. Deng T, Lyon CJ, Bergin S, Caligiuri MA, Hsueh WA. Obesity, inflammation, and cancer. *Annual Review of Pathology* 2016;11:421–449.
52. Whitehead JP, Richards AA, Hickman IJ, Macdonald GA, Prins JB. Adiponectin: A key adipokine in the metabolic syndrome. *Diabetes, Obesity and Metabolism* 2006;8(3):264–280.
53. Arita Y, Kihara S, Ouchi N et al. Paradoxical decrease of an adipose- specific protein, adiponectin, in obesity. *Biochemical and Biophysical Research Communications* 1999;257(1):79–83.
54. Weyer C, Funahashi T, Tanaka S et al. Hypoadiponectinemia in obesity and type 2 diabetes: Close association with insulin resistance and hyperinsulinemia. *The Journal of Clinical Endocrinology and Metabolism* 2001;86(5):1930–1935.
55. Kantartzis K, Fritsche A, Tschritter O et al. The association between plasma adiponectin and insulin sensitivity in humans depends on obesity. *Obesity Research* 2005;13(10):1683–1691.

56. Tschritter O, Fritsche A, Thamer C et al. Plasma adiponectin concentrations predict insulin sensitivity of both glucose and lipid metabolism. *Diabetes* 2003;52(2):239–243.
57. Lindsay RS, Funahashi T, Hanson RL et al. Adiponectin and development of type 2 diabetes in the Pima Indian population. *Lancet* 2002;360(9326):57–58.
58. Yamamoto Y, Hirose H, Saito I, Nishikai K, Saruta T. Adiponectin, an adipocyte-derived protein, predicts future insulin resistance: Two- year follow-up study in Japanese population. *The Journal of Clinical Endocrinology and Metabolism* 2004;89(1):87–90.
59. Snehalatha C, Mukesh B, Simon M, Viswanathan V, Haffner SM, Ramachandran A. Plasma adiponectin is an independent predictor of type 2 diabetes in Asian Indians. *Diabetes Care* 2003;26(12):3226–3229.
60. Spranger J, Kroke A, Mohlig M et al. Adiponectin and protection against type 2 diabetes mellitus. *Lancet* 2003;361(9353):226–228.
61. Greenberg AS, Obin MS. Obesity and the role of adipose tissue in inflammation and metabolism. *The American Journal of Clinical Nutrition* 2006;83(2):461S–465S.
62. Ouchi N, Walsh K. Adiponectin as an anti-inflammatory factor. *Clinica Chimica Acta* 2007;380(12):24–30.
63. Hoenig M, Jordan ET, Glushka J et al. Effect of macronutrients, age, and obesity on 6- and 24-h postprandial glucose metabolism in cats. *American Journal of Physiology: Regulatory, Integrative and Comparative Physiology* 2011;301(6):R1798–R1807.
64. Ellekilde M, Selfjord E, Larsen CS et al. Transfer of gut microbiota from lean and obese mice to antibiotic-treated mice. *Scientific Reports* 2014;4:5922. doi:10.1038/srep05922.
65. Ishioka K, Omachi A, Sasaki N, Kimura K, Saito M. Feline adiponectin: Molecular structures and plasma concentrations in obese cats. *The Journal of Veterinary Medical Science* 2009;71(2):189–5194.
66. Muranaka S, Mori N, Hatano Y et al. Obesity induced changes to plasma adiponectin concentration and cholesterol lipoprotein composition profile in cats. *Research in Veterinary Science* 2011;91(3):358–361.
67. Tvarijonaviciute A, Ceron JJ, Holden SL, Morris PJ, Biourge V, German AJ. Effects of weight loss in obese cats on biochemical analytes related to inflammation and glucose homeostasis. *Domestic Animal Endocrinology* 2012;42(3):129–141.
68. Bjornvad CR, Rand JS, Tan HY et al. Obesity and sex influence insulin resistance and total and multimer adiponectin levels in adult neutered domestic shorthair client-owned cats. *Domestic Animal Endocrinology* 2014;47:55–64. doi:10.1016/j.domaniend.2013.11.006.
69. Coradini M, Rand JS, Morton JM, Arai T, Ishioka K, Rawlings JM. Fat mass, and not diet, has a large effect on postprandial leptin but not on adiponectin concentrations in cats. *Domestic Animal Endocrinology* 2013;45(2):79–88.
70. Prince AL, Chu DM, Seferovic MD, Antony KM, Ma J, Aagaard KM. The perinatal microbiome and pregnancy: Moving beyond the vaginal microbiome. *Cold Spring Harbor Perspectives in Medicine* 2015;5(6). doi:10.1101/cshperspect.a023051.

71. Brunson BL, Zhong Q, Clarke KJ et al. Serum concentrations of adiponectin and characterization of adiponectin protein complexes in dogs. *American Journal of Veterinary Research* 2007;68(1):57–62.
72. Ishioka K, Omachi A, Sagawa M et al. Canine adiponectin: cDNA structure, mRNA expression in adipose tissues and reduced plasma levels in obesity. *Research in Veterinary Science* 2006;80(2):127–132.
73. Bäckhed F, Manchester JK, Semenkovich CF, Gordon JI. Mechanisms underlying the resistance to diet-induced obesity in germ-free mice. *Proceedings of the National Academy of Sciences USA* 2007;104(3):979–984. doi:10.1073/pnas.0605374104.
74. Tvarijonaviciute A, Martínez-Subiela S, Ceron JJ. Validation of 2 commercially available enzyme-linked immunosorbent assays for adiponectin determination in canine serum samples. *Canadian Journal of Veterinary Research* 2010;74(4):279–285.
75. German AJ, Hervera M, Hunter L et al. Improvement in insulin resistance and reduction in plasma inflammatory adipokines after weight loss in obese dogs. *Domestic Animal Endocrinology* 2009;37(4):214–226.
76. Verkest KR, Fleeman LM, Rand JS, Morton JM. Evaluation of beta-cell sensitivity to glucose and first-phase insulin secretion in obese dogs. *American Journal of Veterinary Research* 2011;72(3):357–366.
77. Verkest KR, Rose FJ, Fleeman LM et al. Adiposity and adiponectin in dogs: Investigation of causes of discrepant results between two studies. *Domestic Animal Endocrinology* 2011;41(1):35–41.
78. Alang N, Colleen RK. Weight gain after fecal microbiota transplantation. *Open Forum Infectious Diseases* 2015;2(1). doi:10.1093/ofid/ofv004.
79. Weisberg SP, McCann D, Desai M, Rosenbaum M, Leibel RL, Ferrante AW. Obesity is associated with macrophage accumulation in adipose tissue. *Journal of Clinical Investigation* 2003;112:1796–1808.
80. Cawthorn WP, Sethi JK. TNF-[alpha] and adipocyte biology. *FEBS Letters* 2008;582(1):117–131.
81. Fain JN, Bahouth SW, Madan AK. TNF[alpha] release by the nonfat cells of human adipose tissue. *International Journal of Obesity and Related Metabolic Disorders* 2004;28(4):616–622.
82. Ryden M, Arner P. Tumour necrosis factor-alpha in human adipose tissue – From signalling mechanisms to clinical implications. *Journal of Internal Medicine* 2007;262(4):431–438.
83. Mohamed-Ali V, Goodrick S, Rawesh A et al. Subcutaneous adipose tissue releases interleukin-6, but not tumor necrosis factor-alpha, in vivo. *The Journal of Clinical Endocrinology and Metabolism* 1997;82(12):4196–4200.
84. Hotamisligil GS, Shargill NS, Spiegelman BM. Adipose expression of tumor necrosis factor-alpha: Direct role in obesity-linked insulin resistance. *Science* 1993;259(5091):87–91.
85. Mandalia A, Ward A, Tauxe W, Kraft CS, Dhere T. Fecal transplant is as effective and safe in immunocompromised as non-immunocompromised patients for *Clostridium*

difficile. *International Journal of Colorectal Disease* 2015;September. doi:10.1007/s00384-015-2396-2.
86. Hoenig M, McGoldrick JB, deBeer M, Demacker PNM, Ferguson DC. Activity and tissue-specific expression of lipases and tumor-necrosis factor alpha in lean and obese cats. *Domestic Animal Endocrinology* 2006;30(4):333–344.
87. Stephens JM, Lee J, Pilch PF. Tumor necrosis factor-alpha-insulin resistance in 3T3–L1 adipocytes is accompanied by a loss of insulin receptor substrate-1 and GLUT4 expression without a loss of insulin receptor-mediated signal transduction. *Journal of Biological Chemistry* 1997;272(2):971–976.
88. Qi C, Pekala PH. Tumor necrosis factor-alpha induced insulin resistance in adipocytes. *Proceedings of the Society for Experimental Biology and Medicine* 2000;223(2):128–135.
89. Peraldi P, Xu M, Spiegelman BM. Thiazolidinediones block tumor necrosis factor-alpha induced inhibition of insulin signaling. *Journal of Clinical Investigation* 1997;100(7):1863–1869.
90. Memon RA, Feingold KR, Moser AH, Fuller J, Grunfeld C. Regulation of fatty acid transport protein and fatty acid translocase mRNA levels by endotoxin and cytokines. *American Journal of Physiology* 1998;274:E210–E217.
91. Kita A, Yamasaki H, Kuwahara H et al. Identification of the promoter region required for human adiponectin gene transcription: Association with CCAAT/enhancer binding protein-[beta] and tumor necrosis factor-[alpha]. *Biochemical and Biophysical Research Communications* 2005;331(2):484–490.
92. Kim KY, Kim JK, Jeon JH, Yoon SR, Choi I, Yang Y. c-Jun N-terminal kinase is involved in the suppression of adiponectin expression by TNF- [alpha] in 3T3–L1 adipocytes. *Biochemical and Biophysical Research Communications* 2005;327(2):460–467.
93. Wolin KY, Carson K, Colditz GA. Obesity and cancer. *The Oncologist* 2010;15(6):556–565.
94. Vainio H, Kaaks R, Bianchini F. Weight control and physical activity in cancer prevention: International evaluation of the evidence. *European Journal of Cancer Prevention* 2002;(11 Suppl 2):S94–S100.
95. Janesick A, Blumberg B. Endocrine disrupting chemicals and the developmental programming of adipogenesis and obesity. *Birth Defects Research (Part C)* 2011;93:34–50.
96. Spalding KL, Arner E, Westermark PO et al. Dynamics of fat cell turnover in humans. *Nature* 2008;453:783–787.
97. Ye J, Gimble J. Regulation of stem cell differentiation in adipose tissue by chronic inflammation. *Clinical and Experimental Pharmacology and Physiology* 2011;38:872–878.
98. Assim A, Sallam A, Sallam R. Reactive oxygen species in health and disease. *Journal of Biomedicine and Biotechnology* 2012;2012:1–2.
99. Dikalov SI, Nazarewicz RR. Angiotensin II-induced production of mitochondrial reactive oxygen species: Potential mechanisms and relevance for cardiovascular disease. *Antioxidants & Redox Signaling* 2013;19(10):1085–1094.

100. Bäckhed F, Ding H, Wang T et al. The gut microbiota as an environmental factor that regulates fat storage. *Proceedings of the National Academy of Sciences* 2004;101(44):15718–15723.
101. Ley RE, Turnbaugh PJ, Klein S, Gordon JI. Microbial ecology: Human gut microbes associated with obesity. *Nature* 2006;444(7122):1022–1023. doi:10.1038/4441022a.
102. Turnbaugh PJ, Ley RE, Hamady M, Fraser-Liggett CM, Knight R, Gordon JI. The human microbiome project. *Nature* 2007;449(7164):804–810.
103. Ley RE, Peterson DA, Gordon JI. Ecological and evolutionary forces shaping microbial diversity in the human intestine. *Cell* 2006;124(4):837– 848. doi:10.1016/j.cell.2006.02.017.
104. Bäckhed F, Ley RE, Sonnenburg JL, Peterson DA, Gordon JI. Host-bacterial mutualism in the human intestine. *Science* 2005; 307(5717):1915–1920.
105. Turnbaugh PJ, Ley RE, Mahowald MA, Magrini V, Mardis ER, Gordon JI. An obesity-associated gut microbiome with increased capacity for energy harvest. *Nature* 2006;444(7122):1027–1031.
106. Turnbaugh PJ, Hamady M, Yatsunenko T et al. A core gut microbiome in obese and lean twins. *Nature* 2009;457(7228):480–484.
107. Handl S, German AJ, Holden SL et al. Faecal microbiota in lean and obese dogs. *FEMS Microbiology Ecology* 2013;84(2):332–343. doi:10.1111/1574–6941.12067.
108. Park H-J, Lee S-E, Kim H-B, Isaacson RE, Seo K-W, Song K-H. Association of obesity with serum leptin, adiponectin, and serotonin and gut microflora in Beagle dogs. *Journal of Veterinary Internal Medicine/ American College of Veterinary Internal Medicine* 2015;29(1):43–50. doi:10.1111/jvim.12455.
109. De Filippo C, Cavalieri D, Di Paola M et al. Impact of diet in shaping gut microbiota revealed by a comparative study in children from Europe and Rural Africa. *Proceedings of the National Academy of Sciences* 2010;107(33):14691–14696.
110. Sankar SA, Lagier J-C, Pontarotti P, Raoult D, Fournier P-E. The human gut microbiome, a taxonomic conundrum. *Systematic and Applied Microbiology* 2015;38(4):276–286. doi:10.1016/j.syapm.2015.03.004.
111. Samuel BS, Gordon JI. A humanized gnotobiotic mouse model of host- archaeal-bacterial mutualism. *Proceedings of the National Academy of Sciences* 2006;103(26):10011–10016. doi:10.1073/pnas.0602187103.
112. McNeil NI. The contribution of the large intestine to energy supplies in man. *American Journal of Clinical Nutrition* 1984;39(2):338–342.
113. Li M, Gu D, Xu N et al. Gut carbohydrate metabolism instead of fat metabolism regulated by gut microbes mediates high-fat diet-induced obesity. *Beneficial Microbes* 2014;5(3):335–344. doi:10.3920/ BM2013.0071.
114. Samuel BS, Shaito A, Motoike T et al. Effects of the gut microbiota on host adiposity are modulated by the short-chain fatty-acid binding G protein-coupled receptor, Gpr41. *Proceedings of the National Academy of Sciences* 2008;105(43):16767–16772. doi:10.1073/ pnas.0808567105.

115. Middelbos IS, Boler BMV, Qu A, White BA, Swanson KS, Fahey GC. Phylogenetic characterization of fecal microbial communities of dogs fed diets with or without supplemental dietary fiber using 454 pyrosequencing. *PLOS ONE* 2010;5(3):e9768–e9769. doi:10.1371/journal. pone.0009768.

116. Qiao Y, Sun J, Xia S, Tang X, Shi Y, Le G. Effects of resveratrol on gut microbiota and fat storage in a mouse model with high-fat-induced obesity. *Food & Function* 2014;5(6):1241–1249. doi:10.1039/c3fo60630a.

117. Marques TM, Wall R, O'Sullivan O et al. Dietary trans-10, Cis-12– conjugated linoleic acid alters fatty acid metabolism and microbiota composition in mice. *The British Journal of Nutrition* 2015;113(5):728–738. doi:10.1017/S0007114514004206.

118. Cluny NL, Eller LK, Keenan CM, Reimer RA, Sharkey KA. Interactive effects of oligofructose and obesity predisposition on gut hormones and microbiota in diet-induced obese rats. *Obesity (Silver Spring, MD)* 2015;23(4):769–778. doi:10.1002/oby.21017.

119. Duca FA, Sakar Y, Lepage P et al. Replication of obesity and associated signaling pathways through transfer of microbiota from obese-prone rats. *Diabetes* 2014;63(5):1624–1636. doi:10.2337/db13–1526.

120. Jewell DE, Toll PW, Azain MJ, Lewis RD, Edwards GL. Fiber but not conjugated linoleic acid influences adiposity in dogs. *Veterinary Therapeutics* 2006;7(2):78–85.

121. Grześkowiak Ł, Endo A, Beasley S, Salminen S. Microbiota and probiotics in canine and feline welfare. *Anaerobe* 2015;34(August):14–23. doi:10.1016/j.anaerobe.2015.04.002.

122. Silva BC, Jung LRC, Sandes SHC et al. In vitro assessment of functional properties of lactic acid bacteria isolated from faecal microbiota of healthy dogs for potential use as probiotics. *Beneficial Microbes* 2013;4(3):267–275. doi:10.3920/BM2012.0048.

123. Garcia-Mazcorro JF, Lanerie DJ, Dowd SE et al. Effect of a multi-species synbiotic formulation on fecal bacterial microbiota of healthy cats and dogs as evaluated by pyrosequencing. *FEMS Microbiology Ecology* 2011;78(3):542–554. doi:10.1111/j.1574–6941.2011.01185.x.

124. Luoto R, Kalliomäki M, Laitinen K, Isolauri E. The impact of perinatal probiotic intervention on the development of overweight and obesity: Follow-up study from birth to 10 years. *International Journal of Obesity* 2010;34(10):1531–1537. doi:10.1038/ijo.2010.50.

125. Luoto R, Laitinen K, Nermes M, Isolauri E. Impact of maternal probiotic-supplemented dietary counselling on pregnancy outcome and prenatal and postnatal growth: A double-blind, placebo-controlled study. *The British Journal of Nutrition* 2010;103(12):1792–1799. doi:10.1017/ S0007114509993898.

126. Cengiz Ö, Köksal BH, Tatlı O et al. Effect of dietary probiotic and high stocking density on the performance, carcass yield, gut microflora, and stress indicators of broilers. *Poultry Science* 2015;94(10):2395–2403. doi:10.3382/ps/pev194.

127. Zbinden A, Zbinden R, Berger C, Arlettaz R. Case series of *Bifidobacterium longum* bacteremia in three preterm infants on probiotic therapy. *Neonatology* 2015;107(1):56–59. doi:10.1159/000367985.

128. Eren Z, Gurol Y, Sonmezoglu M, Eren HS, Celik G, Kantarci G. Saccharomyces cerevisiae fungemia in an elderly patient following probiotic treatment. *Mikrobiyoloji Bülteni* 2014;48(2):351–355.
129. Suchodolski JS, Dowd SE, Westermarck E et al. The effect of the macrolide antibiotic tylosin on microbial diversity in the canine small intestine as demonstrated by massive parallel 16S rRNA gene sequencing. *BMC Microbiology* 2009;9:210. doi:10.1186/1471-2180-9-210.
130. Procter TD, Pearl DL, Finley RL et al. A cross-sectional study examining the prevalence and risk factors for anti-microbial-resistant generic escherichia coli in domestic dogs that frequent dog parks in three cities in South-Western Ontario, Canada. *Zoonoses and Public Health* 2014;61(4):250–259. doi:10.1111/zph.12064.
131. Peter D, Sørensen AH, Guardabassi L. Monitoring of antimicrobial resistance in healthy dogs: First report of canine ampicillin-resistant enterococcus faecium clonal complex 17. *Veterinary Microbiology* 2008;132(1-2):190–196. doi:10.1016/j.vetmic.2008.04.026.
132. Eiseman B, Silen W, Bascom GS, Kauvar AJ. Fecal enema as an adjunct in the treatment of pseudomembranous enterocolitis. *Surgery* 1958;44(5):854–859.
133. Brandt LJ. Fecal microbiota transplant: Respice, adspice, prospice. *Journal of Clinical Gastroenterology* 2015;49(Suppl 1 December):S65–S68. doi:10.1097/MCG.0000000000000346.
134. Borody T, Fischer M, Mitchell S, Campbell J. Fecal microbiota transplantation in gastrointestinal disease: 2015 update and the road ahead. *Expert Review of Gastroenterology & Hepatology*. 2015;September:1–13. doi:10.1586/17474124.2015.1086267.

3 肥胖症的病理生理学:合并症和麻醉的注意事项

3.1 引言

大量研究结果表明体重与很多合并症密切相关。本章综述了犬猫体重与合并症之间关系的研究结果,并为犬猫日常管理提供建议。此外,本文还将探讨肥胖宠物围麻醉期(perianesthetic)的管理。

3.2 肥胖症和寿命

一项针对来自7窝、24对拉布拉多犬幼崽一生的研究发现,限制能量摄入组犬摄入的能量比对照组少25%[1]。结果表明,限制能量摄入组的犬体况评分(body condition score,BCS)维持在4~5分,而对照组的犬维持在6~7分(备注:使用9分体况评分系统)。与对照组相比,长期限制能量摄入和维持在较低体况评分犬的平均寿命延长了1.8年($P<0.01$)[1]。采用正交偏最小二乘判别分析(orthogonal projections to latent structures discriminant analysis,OPLS-DA)方法判定2组的代谢表现型,结果表明,2组间存在少数显著差异(如能量和脂蛋白代谢的差异),并且支持了以下观点,即限制能量摄入可降低老龄动物常见合并症(如肥胖症和糖尿病)的发生风险[2]。

在另外一个项目中,针对涵盖了美国养犬俱乐部(AKC)品种目录中的77个品种的718只犬开展研究,研究者评估分析了犬的体长和体重对寿命的影响[3]。结果表明,犬的寿命与体重和体长呈负相关(相关系数r分别为-0.68和-0.60,$P<0.05$)。回归分析结果表明,在健康犬的群体中,体长和体重上的差异可以解释寿命差异的58.5%。最终,作者得出结论,同一品种中体重较轻的犬通常比大体格的犬寿命更长。然而这项研究的一个缺憾是没有评估体况评分。到目前为止,关于猫体重与寿命之间关系的研究仍鲜有报道。总的来说,对犬的研究表明,对患病宠物的常规预防护理应该包括预防超重和肥胖症。

3.3 肥胖症悖论

尽管肥胖症会使人和动物更易患上某些疾病,然而诸多研究却发现了"肥胖症悖论"的存在,即"在慢性疾病确诊后,肥胖症患者的存活率反而更高"。这种悖论已经在不同情况的病人身上得到验证[4-7]。虽然大多数关于人类肥胖症悖论的研究表明体重与存活率之间呈线性关系[8,9],但是也有些研究表明两者呈 U 形或 J 形曲线关系,在这些研究中,体重最大的患者的存活率较低[10,11]。

在兽医学中,肥胖症悖论已经在患有癌症[12]、心力衰竭[13,14]和慢性肾脏病(chronic kidney disease,CKD)的犬猫上得到了验证[15,16]。当患有心力衰竭时,体重增加的犬的存活时间比体重减轻或维持体重的犬更长[13]。当患有慢性肾脏病时,体况评分偏瘦(BCS ≤ 3/9)的犬存活时间显著低于体况评分偏胖的犬(BCS ≥ 4)[15]。此外,当猫被诊断患有慢性肾脏病后,体重大于或等于 4.2 kg 的猫,其存活时间显著长于体重小于 4.2 kg 的猫[16]。

当患有癌症时,猫的体重和体况评分都与存活率显著相关[12]。当患有心力衰竭时,比起轻体重的猫,体重适中猫的存活时间更长。然而,最大体重猫的存活时间比适中体重的猫低,该结果表明,患病猫的体重与患病后存活时间存在 U 形相关关系[4]。在另一项患有慢性肾脏病猫上的研究也发现了类似 J 形曲线关系(图 3.1)[16]。

超重和肥胖症在不同疾病中发挥保护作用的机制尚未完全清楚,但可能通过以下几种假说来解释。这种保护作用可能是多因素的,包括脂肪因子谱(脂肪组织激素)的改变、神经激素的改变和治疗并发疾病的医疗干预。与瘦的人相比,肥胖症患者不但拥有更多的脂肪组织,而且还拥有更多的瘦肉组织。所以,肥胖症患者通常拥有更多的肌肉总量,且极少发生恶病质(cachexia,指因饥饿或疾病造成严重人体耗竭的状态),这是对肥胖症悖论的一个重要解释[7]。肌肉损失对各种疾病的负面影响已有诸多报道[17]。因此,更多的肌肉组织可以为慢性肾脏病或其他分解代谢性疾病的肥胖症患者提供肌肉储备。

总而言之,在治疗患有合并症的超重动物时,在做出是否需要减肥的决定之前,必须考虑肥胖症悖论的存在,如果决定减肥,应该注明减到多少体重合适。例如,对于 1 只被诊断为慢性瓣膜病继发的充血性心力衰竭的犬,当其体况评分(BCS)为 6 分时(9 分制体况评分系统),应该维持其当前体重,而不应该推荐患犬减肥到体况评分 4~5 分(9 分制体况评分系统)。

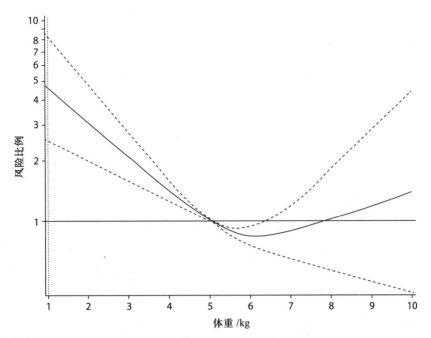

图3.1 569只猫的体重与患有慢性肾脏病风险比例的曲线关系图。虚线表示95%的置信区间。患有慢性肾脏病的猫,体重最轻和最重的猫存活时间均低于体重适中的猫($P<0.0001$)
(改编自Freeman L M et al., *Journal of Veterinary Internal Medicine*, 2016; 30: 1661-1666)

3.4 内分泌疾病

3.4.1 糖尿病

糖尿病(diabetes mellitus, DM)是一种常见的影响犬猫健康的内分泌疾病。犬的糖尿病更像是人类的1型糖尿病,由于胰岛β细胞遭到破坏,导致胰岛素绝对缺乏。大多数猫的糖尿病更像是人类的2型糖尿病,患猫首先在肝脏、肌肉和脂肪组织中发生大面积的胰岛素抵抗,进一步发展为胰岛β细胞衰竭和胰岛素依赖。

3.4.2 犬糖尿病

来自英国全科宠物医院的学者首次提出犬的超重与糖尿病呈正相关[18]。然而,研究人员尚未发现肥胖症与犬糖尿病之间存在直接因果关系的证据。研究发现,拉布拉多犬的阿黑皮素原(proopiomelanocortin, POMC)基因缺失与食欲增加和肥胖症风险增加相关[19]。诸多研究人员尝试寻找阿黑皮素原基因敲除和糖尿病之间的潜在联系,然而并未在犬上发现这种关联[20]。

肥胖症给患糖尿病犬的管理带来了一个挑战。犬的肥胖症会导致胰岛素抵抗。虽然还没有证据表明胰岛素抵抗会导致2型糖尿病,但是这将会影响犬的血糖控制[21]。瘦犬的胰岛素敏感性比超重犬高58%[2]。此外,当犬体重增加接近43%时,其体内基础胰岛素血症和胰岛素抵抗显著增加[22]。肥胖症会进一步影响胰岛素抵抗和血糖控制,导致慢性低度炎症状态。超重的犬减肥后,体内胰岛素抵抗和炎症标记物的浓度也随之降低,特别是肿瘤坏死因子-α(tumor necrosis factor-α,TNF-α),结合珠蛋白(haptoglobin)和C反应蛋白的浓度[23]。良好的血糖控制的一个标志是保持体重。虽然准确评价糖尿病的调节效果很重要,但处于肥胖状态的患犬应制定饮食计划,配合适当的医疗管理,以帮助其控制体重。

3.4.3 猫糖尿病

肥胖是猫患糖尿病最重要的危险因素。患有肥胖症的猫患糖尿病的概率是理想体况评分猫的3.9倍,进而导致胰岛素敏感性降低[24]。瘦猫的胰岛素抵抗发生率仅为10%[25]。猫每超重1 kg,胰岛素敏感性就会降低30%[26]。家养短毛猫糖尿病的发展与黑皮质素受体4(melanocortin 4 receptor,MC4R)基因多态性有关,该基因在能量平衡和食欲调节中发挥作用[27]。脂肪组织会产生脂肪因子,从而影响能量平衡和葡萄糖代谢。其中脂肪因子脂联素与脂肪总量呈负相关,并可以增加胰岛素敏感性[26,28]。但亦有研究未能检测到脂联素浓度在瘦猫、超重猫和肥胖猫之间的显著差异[29]。Zapata等发现,患糖尿病猫的脂联素浓度比瘦猫和超重猫要低[30]。此外,该研究还发现,与瘦猫相比,患糖尿病的猫和超重猫体内的瘦素(一种与脂肪总量呈正相关的脂肪因子)浓度升高。

猫摄入膨化干粮与肥胖症和/或糖尿病的发展有关[31,32]。猫常见的膨化干粮通常比普通日粮的碳水化合物含量高[33]。由于研究性质的不同,以上研究结果不足以判断日粮碳水化合物的摄入量与肥胖症和糖尿病之间是否有因果关系。因为,这些研究存在诸多局限,例如,回忆偏倚、猫主人自己对猫进行体况评分、缺乏常规营养素和日粮摄入量等信息,导致研究结论混淆。日粮中的碳水化合物对猫肥胖症和糖尿病的作用超出了本章节讨论的范围,但是读者如果感兴趣,可以参考Verbrugghe和Hesta的一篇综述进一步探讨[34]。与犬相比,导致猫的肥胖症和糖尿病的原因更可能是过量的能量摄入和自由采食干粮。

虽然日粮中低碳水化合物(low-carbohydrate)在糖尿病发展过程中的作用尚不清楚,但有证据表明低碳水化合物日粮可治疗糖尿病[35-37]。与饲喂中碳水化合物(moderate-carbohydrate)、高纤维日粮的猫相比,饲喂低碳水化合物、低纤维(low-fiber)日粮的猫可以更好地控制血糖和缓解糖尿病[35]。这2种试验日粮含有不同的原料和常规营养素(如蛋白质和脂肪),这种差别影响了试验结果。当猫从高纤维日粮转向低碳水化合物和高蛋白质日粮时,猫对胰岛素的需求降低[37]。对肥胖糖尿病患者的饮食建议还应包括减肥建议,目的是改善外周胰岛素敏感性并有助于缓解糖尿病。有研究显示,减肥可以使猫的胰岛素敏感性恢复正常[26]。

3.4.4 肾上腺皮质功能亢进

肾上腺皮质功能亢进(hyperadrenocorticism,HAC)是中老年犬常见的内分泌疾病。这种病的典型临床表现为内脏型肥胖和腹部膨胀。内脏型肥胖和肾上腺皮质功能亢进的相关性可能与脂肪因子和激素分泌的变化有关。垂体依赖性肾上腺皮质机能亢进(pituitary-dependent hyperadrenocorticism,PDH)的超重犬,其体内的瘦素和胰岛素浓度高于其他超重犬[38]。与此同时,瘦素浓度与血清皮质醇浓度呈线性相关关系。有人研究表明,高瘦素血症的促炎、促血栓生成和促氧化作用可能在肾上腺皮质功能亢进的发展和形成过程中发挥一定作用,然而,这种关联在犬上是否存在,还需要进一步研究[39]。使用曲洛司坦(trilostane)治疗后,犬的胰岛素和瘦素浓度降低,但仍然高于对照组,这可能是间歇性皮质醇增多症和内脏的脂肪含量高于对照组的缘故[38]。在垂体依赖性肾上腺机能亢进患病犬体内,脂联素的变化规律在不同研究中的结果并不一致,一项研究检测到脂联素浓度降低,而另一项研究得出了不同的结果,与拥有相似体况评分的对照组相比,患病犬体内的脂联素浓度没有差异[38,40]。后者未能检测到显著差异的原因可能是由于样本量过少。

使用曲洛司坦治疗肾上腺皮质功能亢进(HAC)可以显著降低超重犬的体重和体况评分[38]。本项研究未报道日粮改变或刻意的体重管理对犬有何影响。除了通过减少肾上腺皮质功能亢进犬的内脏脂肪和潜在改善的医疗管理之外,还未见针对犬减肥管理的研究报道。除了内脏肥胖以外,高脂血症在肾上腺皮质功能亢进患犬中也尤为常见,因此改变饮食有助于控制这种疾病。

3.4.5 甲状腺功能减退

甲状腺激素是全身所有正常代谢功能所必需的,甲状腺功能减退(hypothyroidism)几乎可影响所有器官系统的代谢功能。与甲状腺功能减退相关的代谢紊乱疾病的临床症状,包括嗜睡、运动不耐受,以及在食欲和食物摄入量没有增加情况下的体重增加。与治疗后康复的正常犬相比,甲状腺功能减退犬的能量消耗减少了15%[41]。针对甲状腺功能减退犬的减肥管理,首先应注意恢复其甲状腺的正常功能,以提高其能量消耗。如果仅靠恢复甲状腺激素的正常分泌,却没能达到理想的减肥效果时,推荐采取进一步的营养管理措施。

3.4.6 高脂血症

犬猫高脂血症是指其血浆或血清中胆固醇和/或甘油三酯的浓度高于正常水

平的情况。空腹高脂血症的病因可能是脂肪代谢的原发病变,也可能是继发于其他疾病,其中后者更为常见。肥胖症是高脂血症的重要继发性原因。与瘦犬相比,肥胖犬的胆固醇和/或甘油三酯浓度更高[22,42,43]。与瘦猫相比,肥胖猫的血浆胆固醇和甘油三酯浓度也显著升高[44]。在最近的一项研究中,与瘦猫相比,有主人的、未患糖尿病的超重和肥胖猫的血清甘油三酯浓度更高,但胆固醇浓度不高[45]。在肥胖的比格犬中,将成年犬维持日粮转变为减少能量摄入的低能量密度日粮,血浆胆固醇和甘油三酯浓度在体重显著下降之前即可得到改善[43]。当通过改变饮食的方式减肥时,犬的胆固醇和甘油三酯浓度亦可得到改善[46,47]。

尽早识别犬的空腹高脂血症非常重要,因为该疾病与诸多临床疾病相关,如胰腺炎、肝胆疾病、眼病、胰岛素抵抗和动脉粥样硬化[48]。肥胖猫的血脂异常也与胰岛素抵抗和糖尿病有关[44,45,49,50]。针对继发性高脂血症的治疗应侧重于对基础病情的解决或控制,因此本文推荐高脂血症的肥胖患病宠物采取减肥的做法。高脂血症患者应考虑减少脂肪摄入,特别是当患有严重的、持续或原发性的高脂血症时。在犬和猫的研究中,关于日粮脂肪水平如何降低高脂血症这类研究仍然缺乏。减少脂肪摄入量可能与患病犬猫当前的脂肪摄入量密切相关。对于犬来说,作者建议选取脂肪含量少于 25 g/1 000 kCal ME 的食物。

3.5 心血管和呼吸系统疾病

在肥胖状态下,心血管和呼吸道生理会发生变化。有关病理生理学的进一步讨论,请参阅第 3.9 节。

3.5.1 充血性心力衰竭和心血管疾病

充血性心力衰竭(congestive heart failure,CHF)通常与一种导致瘦肉组织丧失的分解代谢状态相关,该分解代谢状态会促使心脏恶病质发生。许多充血性心力衰竭患者存在超重或肥胖的问题。患充血性心力衰竭的犬和猫中,41% 的犬和 37% 的猫存在超重或者肥胖的问题[13,14]。一项研究表明,在患有充血性心力衰竭的犬中,瘦素水平显著升高[51]。虽然该研究未涉及肥胖犬(体况评分 ≥ 8/9),但相关性分析的结果显示瘦素浓度与体况评分没有相关性。由于瘦素可增加代谢率,并与促炎性细胞因子和儿茶酚胺相互影响,其浓度的升高可能导致心脏恶病质的发生。高于理想体况评分的偏胖患者,在发病后生存时间可能更长,该现象被称为"肥胖症悖论"(见 3.3 节)。针对患有充血性心力衰竭犬的一项研究,未能证明存活时间与体况评分有显著相关性;然而,体重变化与犬的存活时间有显著的相关性,体重增加的犬存活的时间最长[13]。在患有充血性心力衰竭的猫上进行的一项类似研究发现,体重与存活时间呈 U 形关系,体重最低和体重最高的猫存活时间较短[14]。

与人类不同,犬和猫的肥胖症与冠状动脉疾病之间并没有相关性[52]。犬和猫缺乏具有活性的胆固醇酯转运蛋白(cholesterol ester transfer protein,CETP)酶,该酶负责将胆固醇酯从高密度脂蛋白(high-density lipoproteins,HDL)转化为极低密度脂蛋白(very low-density lipoproteins,VLDL)和低密度脂蛋白(low-density lipoproteins,LDL)。这导致胆固醇酯通过高密度脂蛋白分子而非外周组织转运到肝脏。

3.5.2　高血压

犬的肥胖症与高血压(hypertension)之间是否存在相关性尚无定论。虽然有些研究证明肥胖症可能引起高血压[53,54],但是这些发现不具有普遍性。在一项研究中发现,体况评分与高血压没有关联,作者认为肥胖的犬更可能引起高血压合并症(如:心脏病、内分泌疾病)[55]。大多数研究者认为,即使存在这种关联性,通常也是轻微的,不需要特殊的抗高血压治疗[56,57]。猫的肥胖症和高血压之间的关联尚未见报道[49]。

3.5.3　气管塌陷

许多出现气管塌陷(collapsing trachea)的宠物存在超重或肥胖的问题。被诊断为气管塌陷的犬,其体况评分(BCS)的中值为 7/9,其中 64% 的患病犬有超重或肥胖问题[58]。另一项研究对比了 2 种治疗气管塌陷方法的效果,研究者将 103 只犬分为 2 个处理组,每组患病犬的体况评分的中值为 7/9[59]。肥胖症是引发气管塌陷临床症状和病情发展的次要因素之一[60]。虽然在犬上目前尚无对这一方案的应用效果进行评估的研究,但是在改善气管塌陷临床症状的医疗管理中,减肥是一个重要方法[60]。胸内脂肪组织可降低胸壁顺应性和胸腔活动度,从而降低呼吸功能。通过减少多余的脂肪组织,包括腹腔内和胸腔外的脂肪沉积,可以改善胸壁顺应性。减肥主要应通过饮食管理来实现,因为许多患病宠物如果进行运动,则会加剧临床症状恶化。

3.6　肾脏和泌尿疾病

3.6.1　肾脏疾病

体况评分异常可能是肾脏疾病发展和恶化过程中的一个危险因素。在一项分析慢性肾脏疾病(CKD)诊断过程中相关危险因素的回顾性研究中,研究者发现,比起拥有理想或较高体况评分的猫,体况评分为 1~2 分(5 分制体况评分系统)的猫患病的风险更高[61]。当确诊为慢性肾脏疾病后,体重大于或者等于 4.2 kg 猫的存活时间显著长于体重小于 4.2 kg 的猫[16]。

尽管有证据表明体况评分(BCS)≥4/9 的犬可能有助于延长寿命[15],但肥胖症也会引起肾脏的不良反应。试验诱导的肥胖犬表现为动脉血压升高,肾素-血管紧张素系统激活。这会造成肾小球超滤,并会导致与肾小球损伤一致的组织学病变[62]。肥胖症是否是诱导犬和猫自然发生肾脏疾病的一个重要危险因素还有待进一步研究。

一项研究评估了 37 只肥胖犬减肥对肾功能指标的影响[63]。结果表明,减肥后尿素和尿比重增加。尿蛋白与肌酐(urine protein creatinine,UPC)比值、尿白蛋白与肌酐(urine albumin creatinine,UAC)比值和肌酐均在减肥后下降。3 种新的肾功能生物标记物(同型半胱氨酸、半胱氨酸蛋白酶抑制剂 C 和聚集素)随减肥而下降。作者由此得出结论,犬的肥胖症会导致肾功能发生亚临床改变,而这种亚临床改变可以随着体重减轻而得到改善[63]。与之相反,另一项关于犬的研究发现,尿蛋白与肌酐(UPC)和体况评分(BCS 4~5 分 vs. BCS>6/9 分)没有联系[64]。由于样本量有限,该研究不能区分超重犬和肥胖犬的详细情况。如果一个超重或肥胖的动物有明显的蛋白尿,可能需要考虑使用低蛋白质和低热量的饮食来达到减肥目的,同时不加剧蛋白尿问题。

3.6.2 泌尿疾病

体重增加是猫下泌尿道综合征(feline lower urinary tract disease,FLUTD)的一个危险因素,该综合征包括猫特发性膀胱炎(feline idiopathic cystitis,FIC,猫无明确病因的膀胱炎)、尿石症(urolithiasis)、尿路感染(urinary tract infections,UTI)、尿道梗阻(urethral obstruction,UO)和肿瘤[65-67]。体重与泌尿疾病之间的关联在猫特发性膀胱炎(FIC)病例中的报道最多[65,66],这可能与另一项研究的发现有关,即低活动水平与猫下泌尿道综合征的发展相关[68]。然而,另一项流行病学研究没有发现体重与猫下泌尿道综合征有相关性[69]。与对照组相比,猫的体重增加是诱发尿道阻塞的一个危险因素[70]。

已有研究报道了犬的体况评分与草酸钙(calcium oxalate,CaOx)尿石症(urolithiasis)的关系。一项研究发现,草酸钙尿石症患病犬的体况评分的中值(6/9,范围:4~7/9)显著高于其对照组(5/9,范围:4~7/9);然而,该研究中 2 组犬之间超重犬的比例没有显著差异,均不包含任何肥胖犬(BCS≥8/9)[71]。尽管诸多研究报道了肥胖症和人类女性尿失禁之间的关联性[72],但在兽医学上还没有这种关联性的报道。在病态肥胖犬(体脂>45%,利用双能吸收法评估)中,已有报道指出无症状菌尿的患病率增加[73]。

3.7 骨科和神经内科疾病

3.7.1 骨科疾病

根据理论上的推测,肥胖症是形成骨关节炎(osteoarthritis,OA)的一个危险因素。体脂肪组织分泌的炎性介质可能诱发骨科疾病的发生和发展[74,75]。一项针对92只体成分不同的拉布拉多犬的研究表明,空腹血浆白细胞介素-6(interleukin-6,IL-6)和单核细胞趋化蛋白-1(monocyte chemotactic protein-1,MCP-1)的浓度均与体况评分的增加相关,这凸显了肥胖症和慢性炎症之间的关系[76]。一项通过日粮诱导肥胖大鼠的研究表明,骨关节炎与较高的体脂肪含量相关,但是不一定与体重相关[77]。

关于犬的体重与退变性关节病(DJD)和骨性关节炎之间的关系,已有诸多报道。最早记录这种关系的研究表明,肥胖症与关节和/或运动问题有关。虽然该研究未能采用客观的方法来记录骨关节炎(OA),但是却为以后的研究打下了基础[78]。在一项对48只拉布拉多犬的终生研究中,处理组比对照组热量摄入减少25%,结果却发现肘关节炎的发生率没有差异[79]。对照组的犬在6岁时肘关节骨性关节炎的严重程度显著高于处理组,但是在8岁和死亡时2组间均无显著差异。在同一试验中,研究者使用放射学检测的手段对受试犬检查,结果发现热量限制与延迟髋关节骨性关节炎的发展相关。热量限制处理组的犬检测到骨性关节炎的中位年龄是12岁,而对照组的中位年龄只有6岁[80]。在另一项研究中,对2岁以上的4个大型犬种(金毛犬、拉布拉多犬、德国牧羊犬、罗威纳犬)的15 742只犬进行X射线检测发现,犬的体重是退变性关节病的重要危险因素。值得注意的是,这些研究中没有报道体况评分的数据[81]。

犬的颅交叉韧带断裂也可能与肥胖症有关。一项研究表明,双侧颅内交叉韧带(cranial cruciate ligament,CCL)断裂犬比单侧颅内交叉韧带断裂犬的体重大,但在统计学上不显著。该研究并没有记录体况评分[82]。另一项研究表明,与拥有理想体重的犬相比,肥胖的犬患颅内交叉韧带断裂的风险增加了近4倍。作者认为肥胖与颅内交叉韧带断裂之间的相关性是由于四肢的负荷增加和关节内韧带的张力所致[83]。

也许这并不奇怪,减肥是骨性关节炎最有效的治疗方法之一。诸多研究证明,减肥可以改善患有骨性关节炎的超重犬的运动能力[84-86]。体重减少6.1%以上即可显著降低犬的跛行率[86]。在猫上,虽然体重和体况评分不被普遍认为是发生骨科疾病的危险因素[87],但一项研究表明,超重的猫出现跛足的可能性是拥有理想体况猫的 2.9 倍[24]。在缅因库恩猫、波斯猫和喜马拉雅猫中，髋关节发育不良的概率明显较高，这表明体尺（身体大小）与骨科疾病之间可能存在联系[88]。

3.7.2 神经内科疾病

患有神经内科疾病的犬和猫与患有骨科疾病的动物面临着类似的运动问题。在患有椎间盘疾病（intervertebral disc disease，IVDD）的腊肠犬中，体重和体况评分都与该疾病风险无关[89,90]。然而，其他体尺检测指标[例如：T1-S1 距离增长，（T1 为第 1 胸椎，S1 为第 1 骶骨），鬐甲突出，骨盆围缩小]与神经功能障碍的严重程度相关[90]。对因椎间盘疾病而接受脊柱减压手术的大群体分析发现，体重大于 20 kg 的犬术后发生椎间盘脊柱炎的风险更高[91]。该研究没有评估体况评分。犬的椎间盘脊柱炎发生率极低（2.2%），其他因素也（如犬种、麻醉时间长短）可能会对结果造成影响。

3.8 肿瘤

肿瘤是一个涉及广泛的疾病类型，在患者的体成分及其与各个疾病患病率和预后的关系方面有诸多发现。在一项对 196 只兰波格犬（Leonberger）从出生到死亡的大规模前瞻性群体研究中，评估了原发性骨肿瘤的患病率。研究结果表明，196 只犬中，有 9 只犬死于原发性骨癌。与其他正常犬相比，这 9 只犬在生长期和刚成年时体重较重，这表明体成分可能对骨肿瘤的发展有影响[92]。在一项对 325 只患有脾脏肿块犬的研究中，体重小于或等于 27.8 kg 的犬血管肉瘤（hemangiosarcoma）诊出率显著低于体重大于 27.8 kg 的犬[93]。超重或肥胖的犬患乳腺癌的风险更高，可能是由于肿瘤-脂肪细胞相互作用和与激素受体相关的肿瘤生长所致[94,95]。患有乳腺肿瘤的超重或肥胖的犬，脂联素表达降低，巨噬细胞数量增加，与预后不良的因素如高组织学分级（high histologic grade）和淋巴管浸润（lymphatic invasion）显著相关[94]。当对体成分进行更深入地分析时，患癌症犬的体况评分呈现比较大的跨度，这可能取决于潜在的肿瘤及其合并症。在一项对 1 777 只患癌症犬的研究中，平均体况评分为 5.3/9 分[96]。另一项研究表明，超重犬（BCS ≥ 7/9 分，占 29%）患癌症的比例显著高于体重偏瘦犬（BCS ≤ 3/9 分，占 4%）[97]。在患有各种肿瘤的猫上，肌肉损失（如恶病质）比那些体况评分低的猫更为常见[12]。

一项对患癌症猫的研究发现，体重大于 5 kg 和/或体况评分大于或者等于 5/9 分，猫的存活率较高[12]。在诊断为淋巴瘤或骨肉瘤的犬中，体况评分较低的犬存活时间较短。治疗期间体重增加与存活率增加相关[98]。

在对患有肥胖症和肿瘤的猫或犬实施减肥计划之前,应对其进行一个完整的个体评估,包括癌症分期(cancer stage)、临床症状和预后。肿瘤患者减肥的一个首要目标应该是提高生活质量。例如,一只因原发性骨肿瘤而截肢的肥胖犬,减肥可能是提高患犬活动能力的必要条件。在更容易诱发恶病质(如胃肠道淋巴瘤)的癌症动物中,只有当动物同时患有超重或肥胖症引起的合并症(如严重的呼吸系统疾病或骨关节炎)时,主动减肥才更可能对患病动物有益。否则,由于肥胖症悖论现象,更明智的做法是让这些动物保持一定程度的超重。

3.9 麻醉注意事项

超重的患病动物可能给麻醉师带来诸多困难,包括肥胖症引起的器官系统功能改变、药物治疗问题和后勤医疗支持的挑战。许多已发表的注意事项和推荐方案都是从人类的治疗经验中推断出来的[99,100],包括医疗资源的利用、手术时间的延长,以及由于皮下氧张力降低而增加手术部位感染的风险。在获得更多的特定物种信息之前,采用相对保守的方法对肥胖宠物进行麻醉管理是很有必要的。

3.9.1 肥胖症引起的生理变化

3.9.1.1 呼吸系统疾病病理生理学

在围手术期,肥胖宠物可能会迅速去饱和(desaturate),并表现出高碳酸血症(hypercapnia)的倾向,有时会出现严重的高碳酸血症[101]。肥胖症可能使插管变得复杂,因为沉积在咽腔的脂肪会使呼吸道可视化变得困难,尤其是短头颅(brachycephalic)的患病宠物。胸壁和腹部的脂肪堆积会影响呼吸机能,进而影响呼吸系统。此外,肺血容量增加和气道关闭导致呼吸道顺应性降低[102]。

随着体重的增加,肺容量减少,主要包括潮气量(tidal volume,VT)和功能余气量(functional residual capacity,FRC)的减少。肺的功能余气量(FRC)呈指数级下降,最终可能接近残气量[103,104]。低肺容量(lung volume)呼吸减少呼气流量,增加呼吸系统阻力,导致气道关闭,并造成通气与灌流差异。肥胖症可导致气道阻力增加和肺顺应性下降,这增加了呼吸做功。在 6 min 步行测试(6-minute walk test,6MWT)中,肥胖犬的步行距离低于瘦犬,并且减肥可提高步行距离[105]。该研究的作者将 6 min 步行测试(6MWT)表现不佳作为肥胖犬生活质量下降的标志,并初步将其归因于与肥胖相关的呼吸功能障碍。

肥胖的猫呼吸时潮气量减少且气流受限[104,106]。当肥胖犬患有重复呼吸(rebreathing)引起的呼吸过度时,功能余气量(FRC)和呼气流量限制均减少[103]。在静止状态下,肥胖和非肥胖犬的吸气或呼气流量没有差异;然而,肥胖的人在意识清醒站立的状态下，肥胖症对气道功能的影响也很小[107]。仰卧位（相当于犬和猫的背侧卧）增加了肥胖的人的呼气流量限制。肥胖比格犬的气道高反应性（hyperreactivity）已被证实，并可能受肥胖症引起的全身炎症状态导致；然而，这种现象并没有在肥胖的猫上发现[106,108]。

呼吸顺应性和肺容积的降低导致通气和灌注分布不均，从而影响呼吸气体交换。肥胖的人在休息时会出现低氧血症（hypoxemia），但一项针对12只意识清醒肥胖犬的小规模研究并未证实这一发现[108]。在一项对深度镇静肥胖犬的临床研究中，证实了低氧血症需要补充氧气。减肥后，动脉血氧分压（pressure of arterial oxygen，PaO_2）得到改善，需要补充的氧气减少，特别是胸腔脂肪沉积减少[使用双能X射线吸收测定法（dual-energy x-ray absorptiometry，DXA）证实][109]。

为了保持静息每分钟通气量（minute ventilation）和避免高碳酸血症，肥胖的人和犬会增加呼吸频率[105,108,109]。然而，由于呼吸顺应性慢性降低和气道阻力增加，对二氧化碳浓度的通气反应下降，并可发展为肥胖低通气综合征。围手术期补充氧气会进一步降低通气动力（ventilatory drive），麻醉医生应做好人工或机械通气的准备[105]。虽然在12只意识清醒的肥胖犬上没有发现高碳酸血症，但是该综合征也尚未在伴侣动物上进行专门的研究[105]。

给所有患病宠物预吸氧是一项推荐的处理方法，这对肥胖动物尤其重要，因为动物在麻醉后血氧会迅速去饱和。呼气末二氧化碳分压监测是一种无创且易于应用的方法，用于监测麻醉期间的通气是否充足。肥胖动物在麻醉期间可能需要人工或机械通气。上气道功能障碍在麻醉后期发生的可能性更大。在麻醉恢复期间，应对肥胖动物密切监测，并且应该考虑在此期间根据情况补充氧气。

3.9.1.2　心血管病理生理学

试验性诱导的肥胖症导致犬的交感神经活动持续增加，静息心率增加，心率变异性降低，最终导致全身性高血压[110]。与瘦犬（BCS：5/9分）或轻微超重的犬（BCS：6~7/9分）相比，肥胖犬（BCS ≥ 8/9分）在休息和运动时心率更高[105]。一项对临床犬群的研究表明，肥胖症不是导致系统性高血压的危险因素之一。然而，该研究使用了5级体况评分系统，而且没有对受试者的评分方式进行准确描述[55]。另一项研究发现，肥胖犬（BCS ≥ 7/9分）的动脉收缩压（systolic arterial blood pressure，SAP）（通过多普勒设备测量）比瘦犬（BCS ≤ 6/9分）高[111]。肥胖犬的收缩期动脉压的平均值 ± 标准偏差为(153 ± 19)mmHg，而瘦犬的收缩期动脉压为(133 ± 20)mmHg。虽然这些数值之间没有显著性差异，但与瘦犬相比，更多的肥胖犬被归类为有中度末端器官损伤的风险。

在肥胖的犬中，已经发现了不同类型的心脏舒张功能障碍，通过超声心动图（echocardiography）证实了心脏舒张功能受损和左心室充盈压力增加。肥胖的犬表现出左心室非对称的向心性肥厚,但是心室无扩张[111]。这些变化的临床意义以及与心肌或麻醉结果的相关性尚不清楚。然而,在为肥胖的犬猫制定围麻醉期管理计划时,应认识到全身性高血压和/或舒张功能减退的可能性。

3.9.1.3 其他器官功能障碍

麻醉师特别感兴趣的是肥胖猫患糖尿病的概率是瘦猫的4倍[24]。患有糖尿病的犬猫在围手术期需要控制血糖浓度。然而,提升手术效果所需的血糖控制浓度尚不清楚。骨关节炎引起的疼痛在肥胖的犬猫上更为常见[112]。为避免加重关节疼痛,治疗过程中应特别注意四肢的放置和填充支撑物。对其他合并症(如气管塌陷和各种内分泌疾病)也需要进行特殊的围麻醉期管理,并且在麻醉前应仔细检查肥胖患病宠物。

3.9.2 肥胖症所致的药理学差异

肥胖症可能影响药代动力学参数。药代动力学涉及药物在体内的吸收、分配、代谢和消除。

3.9.2.1 药物吸收

肥胖症可能在多个方面改变药物的吸收。由于肥胖受试者的皮下血流量较低,这可能会减少或延迟皮下给药的药物吸收[113]。因为皮下沉积大量的脂肪组织,所以注入肌内的药物可能会沉积于皮下。对于某些药物而言,会极大改变药物的吸收特性。例如,对猫皮下注射丁丙诺啡(buprenorphine)时,猫不能有效地吸收该药物(也不会产生明显的镇痛效果),而通过肌内和静脉内注射后,该药物则会呈现典型的吸收和分布特征,而且会增加热痛阈值[114]。

3.9.2.2 药物分布

药物分布容积(distribution volume)是一个药代动力学参数,用于确定药物的载药量。药物分布容积是指体内药量与血药浓度之间的平衡关系,取决于药物的理化性质、组织血流量和血浆蛋白质结合率。体成分的变化可增加或减少药物分布容积,药物分布容积取决于药物是亲脂性还是亲水性。随着肥胖症的发展,瘦体重(lean body weight)与脂肪重量增加,但不是同时增加,甚至可能出现反向关系(即体脂百分比可能大于瘦体重)[115]。大多数麻醉药和镇静剂是亲脂的,因此在肥胖患者中可能有更大的分布容积。肥胖症也会改变组织血流量,由于过多脂肪组织的代谢需求增加,血容量和心输出量会相应增加。此外,心脏功能障碍可能改变组织血流量和药物分布。这些变化可能会导致分布容积增加。在肥胖症患者中,白蛋白与药物的结合似乎没有变化,但α-1酸性糖蛋白(alpha-1 acid glycoprotein)与基础药物的结合情况存在不确定性[116]。

针对人类肥胖患者使用镇静剂和麻醉剂剂量变化的问题,研究者已经给出了诸多建议,包括根据患者理想体重使用亲水性药物剂量,根据患者总体重使用亲脂性药物剂量,以及根据患者体重变化(通常是理想体重加上 20%~40%)来调整药物剂量(即根据脂肪组织和瘦肉组织的分布调整药物剂量)[117]。遗憾的是,很少有数据支撑这些建议,因为大多数药代动力学研究是在正常人群中进行的。在理想情况下,这些医疗建议应该对肥胖受试者进行物种特异性药代动力学研究。到目前为止,只有一项对犬的研究调查发现,由于脂肪增加而引起了麻醉药物剂量的变化。这是一项未收集药代动力学数据的临床研究。以总体重为参考基准,与正常体重对照组[(2.2±0.5)mg/kg][在术前服用美托咪定(medetomidine)和布托啡诺(butorphanol)]相比,超重和肥胖犬需要较低剂量的异丙酚(propofol)[(1.8±0.4)mg/kg][118]。由于缺乏更严格的证据,最好的方法是逐步增加镇静剂、镇痛剂和麻醉剂的初始注射剂量。

3.9.2.3 药物代谢与消除

由于体内药物代谢分解和消除导致药物血浆浓度下降,这个过程被称为"药物清除"。无论是通过反复给药还是输液给药,药物清除率决定了注射药物的维持剂量。血浆药物清除率是所有器官药物清除率的总和,包括肝、肾、肺和其他一般次要途径。在肥胖的人和大鼠上,肝脏代谢的变化随着细胞色素 p450(cytochrome p450,CYP)2E1 酶活性的增加而变化,而 CYP 3A4 对药物的清除率较低[119]。在肥胖的犬或猫上,仍没有药物清除率的物种特异性信息。对于初始加载剂量,应谨慎滴定可注射药物的剂量,以确定其在肥胖动物上的合适维持剂量。

3.10 结论

肥胖症会对寿命产生负面影响,但是当患有癌症、心力衰竭和慢性肾脏病时,超重患病宠物的存活时间反而可能更长。超重或肥胖的犬猫通过减肥,将对许多方面疾病,如内分泌、心血管、呼吸、肾脏、泌尿、骨科、神经系统疾病和肿瘤疾病有益,但是当特定的疾病与肥胖悖论相关时,必须考虑体重减轻的程度。由于肥胖会引起器官系统功能改变、药物治疗问题和后勤支持方面的挑战,因此超重患病宠物在接受麻醉时需要额外注意。

参考文献

1. Kealy RD, Lawler DF, Ballam JM et al. Effects of diet restriction on life span and age-related changes in dogs. *Journal of the American Veterinary Medical Association* 2002;220:1315–1320.

2. Richards SE, Wang Y, Claus SP et al. Metabolic phenotype modulation by caloric restriction in a lifelong dog study. *Journal of Proteome Research* 2013;12:3117–3127.
3. Greer KA, Canterberry SC, Murphy KE. Statistical analysis regarding the effects of height and weight on life span of the domestic dog. *Research in Veterinary Science* 2007;82:208–214.
4. Oga EA, Eseyin OR. The obesity paradox and heart failure: A systematic review of a decade of evidence. *J Obes* 2016;2016:9040248.
5. Carnethon MR, Rasmussen-Torvik LJ, Palaniappan L. The obesity paradox in diabetes. *Current Cardiology Reports* 2014;16:446.
6. Park J, Ahmadi S-F, Streja E et al. Obesity paradox in end-stage kidney disease patients. *Progress in Cardiovascular Diseases* 2014;56:415–425.
7. Kalantar-Zadeh K, Horwich TB, Oreopoulos A et al. Risk factor paradox in wasting diseases. *Current Opinion in Clinical Nutrition and Metabolic Care* 2007;10:433–442.
8. Kovesdy CP, Anderson JE, Kalantar-Zadeh K. Paradoxical association between body mass index and mortality in men with CKD not yet on dialysis. *American Journal of Kidney Diseases* 2007;49:581–591.
9. Fonarow GC, Srikanthan P, Costanzo MR et al. An obesity paradox in acute heart failure: Analysis of body mass index and inhospital mortality for 108,927 patients in the Acute Decompensated Heart Failure National Registry. *American Heart Journal* 2007;153:74–81.
10. Kapoor JR, Heidenreich PA. Obesity and survival in patients with heart failure and preserved systolic function: A U-shaped relationship. *American Heart Journal* 2010;159:75–80.
11. Johnson AP, Parlow JL, Whitehead M et al. Body mass index, outcomes, and mortality following cardiac surgery in Ontario, Canada. *Journal of the American Heart Association* 2015;4:e002140.
12. Baez JL, Michel KE, Sorenmo K et al. A prospective investigation of the prevalence and prognostic significance of weight loss and changes in body condition in feline cancer patients. *Journal of Feline Medicine and Surgery* 2007;9:411–417.
13. Slupe JL, Freeman LM, Rush JE. Association of body weight and body condition with survival in dogs with heart failure. *Journal of Veterinary Internal Medicine* 2008;22:561–565.
14. Finn E, Freeman LM, Rush JE et al. The relationship between body weight, body condition, and survival in cats with heart failure. *Journal of Veterinary Internal Medicine* 2010;24:1369–1374.
15. Parker VJ, Freeman LM. Association between body condition and survival in dogs with acquired chronic kidney disease. *Journal of Veterinary Internal Medicine* 2011;25:1306–1311.
16. Freeman LM, Lachaud MP, Matthews S et al. Evaluation of weight loss over time in cats with chronic kidney disease. *Journal of Veterinary Internal Medicine* 2016;30:1661–1666.

17. Morley JE, Thomas DR, Wilson MM. Cachexia: Pathophysiology and clinical relevance. *American Journal of Clinical Nutrition* 2006;83:735–743.
18. Mattin M, O'Neill D, Church D et al. An epidemiological study of diabetes mellitus in dogs attending first opinion practice in the UK. *The Veterinary Record* 2014;174:349.
19. Raffan E, Dennis RJ, O'Donovan CJ et al. A deletion in the canine POMC gene is associated with weight and appetite in obesity-prone Labrador retriever dogs. *Cell Metabolism* 2016;23:893–900.
20. Davison LJ, Holder A, Catchpole B et al. The canine POMC gene, obesity in Labrador retrievers and susceptibility to diabetes mellitus. *Journal of Veterinary Internal Medicine* 2017;31:343–348.
21. Verkest KR, Fleeman LM, Morton JM et al. Compensation for obesity- induced insulin resistance in dogs: Assessment of the effects of leptin, adiponectin, and glucagon-like peptide-1 using path analysis. *Domestic Animal Endocrinology* 2011;41:24–34.
22. Bailhache E, Nguyen P, Krempf M et al. Lipoproteins abnormalities in obese insulin-resistant dogs. *Metabolism* 2003;52:559–564.
23. German AJ, Hervera M, Hunter L et al. Improvement in insulin resistance and reduction in plasma inflammatory adipokines after weight loss in obese dogs. *Domestic Animal Endocrinology* 2009;37:214–226.
24. Scarlett JM, Donoghue S. Associations between body condition and disease in cats. *Journal of the American Veterinary Medical Association* 1998;212:1725–1731.
25. Hoenig M, Pach N, Thomaseth K et al. Cats differ from other species in their cytokine and antioxidant enzyme response when developing obesity. *Obesity (Silver Spring)* 2013;21:E407–E414.
26. Hoenig M, Thomaseth K, Waldron M et al. Insulin sensitivity, fat distribution, and adipocytokine response to different diets in lean and obese cats before and after weight loss. *American Journal of Physiology: Regulatory, Integrative and Comparative Physiology* 2007;292:R227–R234.
27. Forcada Y, Holder A, Church DB et al. A polymorphism in the melanocortin 4 receptor gene (MC4R:c.92C>T) is associated with diabetes mellitus in overweight domestic shorthaired cats. *Journal of Veterinary Internal Medicine* 2014;28:458–464.
28. Ishioka K, Omachi A, Sasaki N et al. Feline adiponectin: Molecular structures and plasma concentrations in obese cats. *Journal of Veterinary Medical Science* 2009;71:189–194.
29. Bjornvad CR, Rand JS, Tan HY et al. Obesity and sex influence insulin resistance and total and multimer adiponectin levels in adult neutered domestic shorthair client-owned cats. *Domestic Animal Endocrinology* 2014;47:55–64.
30. Zapata RC, Meachem MD, Cardoso NC et al. Differential circulating concentrations of adipokines, glucagon and adropin in a clinical population of lean, overweight and diabetic cats. *BMC Veterinary Research* 2017;13:85.
31. Öhlund M, Egenvall A, Fall T et al. Environmental risk factors for diabetes mellitus in cats. *Journal of Veterinary Internal Medicine* 2017;31:29–35.

32. Rowe E, Browne W, Casey R et al. Risk factors identified for owner- reported feline obesity at around one year of age: Dry diet and indoor lifestyle. *Preventive Veterinary Medicine* 2015;121:273–281.
33. Plantinga EA, Bosch G, Hendriks WH. Estimation of the dietary nutrient profile of free-roaming feral cats: Possible implications for nutrition of domestic cats. *British Journal of Nutrition* 2011;106:S35–S48.
34. Verbrugghe A, Hesta M. Cats and carbohydrates: The carnivore fantasy? *Vet Sci* 2017;4:E55.
35. Bennett N, Greco DS, Peterson ME et al. Comparison of a low carbohydrate-low fiber diet and a moderate carbohydrate-high fiber diet in the management of feline diabetes mellitus. *Journal of Feline Medicine and Surgery* 2006;8:73–84.
36. Mazzaferro EM, Greco DS, Turner AS et al. Treatment of feline diabetes mellitus using an alpha-glucosidase inhibitor and a low-carbohydrate diet. *Journal of Feline Medicine and Surgery* 2003;5:183–189.
37. Frank G, Anderson W, Pazak H et al. Use of a high-protein diet in the management of feline diabetes mellitus. *Veterinary Therapeutics: Research in Applied Veterinary Medicine* 2001;2:238–246.
38. Cho K-D, Paek J, Kang J-H et al. Serum adipokine concentrations in dogs with naturally occurring pituitary-dependent hyperadrenocorticism. *Journal of Veterinary Internal Medicine* 2014;28:429–436.
39. Katagiri H, Yamada T, Oka Y. Adiposity and cardiovascular disorders: Disturbance of the regulatory system consisting of humoral and neuronal signals. *Circulation Research* 2007;101:27–39.
40. Cabrera Blatter MF, del Prado B, Miceli DD et al. Interleukin-6 and insulin increase and nitric oxide and adiponectin decrease in blind dogs with pituitary-dependent hyperadrenocorticism. *Research in Veterinary Science* 2012;93:1195–1202.
41. Greco DS, Rosychuk RA, Ogilvie GK et al. The effect of levothyroxine treatment on resting energy expenditure of hypothyroid dogs. *Journal of Veterinary Internal Medicine* 1998;12:7–10.
42. Chikamune T, Katamoto H, Ohashi F et al. Serum lipid and lipoprotein concentrations in obese dogs. *Journal of Veterinary Medical Science* 1995;57:595–598.
43. Jeusette IC, Lhoest ET, Istasse LP et al. Influence of obesity on plasma lipid and lipoprotein concentrations in dogs. *American Journal of Veterinary Research* 2005;66:81–86.
44. Hoenig M, Wilkins C, Holson JC et al. Effects of obesity on lipid profiles in neutered male and female cats. *American Journal of Veterinary Research* 2003;64:299–303.
45. Hoenig M, Traas AM, Schaeffer DJ. Evaluation of routine hematology profile results and fructosamine, thyroxine, insulin, and proinsulin concentrations in lean, overweight, obese, and diabetic cats. *Journal of the American Veterinary Medical Association* 2013;243:1302–1309.

46. Diez M, Michaux C, Jeusette I et al. Evolution of blood parameters during weight loss in experimental obese Beagle dogs. *Journal of Animal Physiology and Animal Nutrition* 2004;88:166–171.
47. Peña C, Suarez L, Bautista-Castaño I et al. Effects of low-fat high-fibre diet and mitratapide on body weight reduction, blood pressure and metabolic parameters in obese dogs. *Journal of Veterinary Medical Science* 2014;76:1305–1308.
48. Xenoulis PG, Steiner JM. Canine hyperlipidaemia. *Journal of Small Animal Practice* 2015;56:595–605.
49. Jordan E, Kley S, Le N-A et al. Dyslipidemia in obese cats. *Domestic Animal Endocrinology* 2008;35:290–299.
50. Nishii N, Maeda H, Murahata Y et al. Experimental hyperlipemia induces insulin resistance in cats. *Journal of Veterinary Medical Science* 2012;74:267–269.
51. Fonfara S, Hetzel U, Tew SR et al. Leptin expression in dogs with cardiac disease and congestive heart failure. *Journal of Veterinary Internal Medicine* 2011;25:1017–1024.
52. Jahangir E, De Schutter A, Lavie CJ. The relationship between obesity and coronary artery disease. *Translational Research: The Journal of Laboratory and Clinical Medicine* 2014;164:336–344.
53. Rocchini AP, Moorehead C, Wentz E et al. Obesity-induced hypertension in the dog. *Hypertension* 1987;9:III64–II68.
54. Granger JP, West D, Scott J. Abnormal pressure natriuresis in the dog model of obesity-induced hypertension. *Hypertension* 1994;23:I8–I11.
55. Pérez-Sánchez AP, Del-Angel-Caraza J, Quijano-Hernández IA et al. Obesity-hypertension and its relation to other diseases in dogs. *Veterinary Research Communications* 2015;39:45–51.
56. Bodey AR, Michell AR. Epidemiological study of blood pressure in domestic dogs. *Journal of Small Animal Practice* 1996;37:116–125.
57. Montoya JA, Morris PJ, Bautista I et al. Hypertension: A risk factor associated with weight status in dogs. *Journal of Nutrition* 2006;136:2011S–2013S.
58. Rubin JA, Holt DE, Reetz JA et al. Signalment, clinical presentation, concurrent diseases, and diagnostic findings in 28 dogs with dynamic pharyngeal collapse (2008–2013). *Journal of Veterinary Internal Medicine* 2015;29:815–821.
59. Tinga S, Thieman Mankin KM, Peycke LE et al. Comparison of outcome after use of extra-luminal rings and intra-luminal stents for treatment of tracheal collapse in dogs. *Veterinary Surgery* 2015;44:858–865.
60. Maggiore AD. Tracheal and airway collapse in dogs. *The Veterinary Clinics of North America: Small Animal Practice* 2014;44:117–127.
61. Greene JP, Lefebvre SL, Wang M et al. Risk factors associated with the development of chronic kidney disease in cats evaluated at primary care veterinary hospitals. *Journal of the American Veterinary Medical Association* 2014;244:320–327.
62. Henegar JR, Bigler SA, Henegar LK et al. Functional and structural changes in the kidney in the early stages of obesity. *Journal of the American Society of Nephrology*

2001;12:1211–1217.
63. Tvarijonaviciute A, Ceron JJ, Holden SL et al. Effect of weight loss in obese dogs on indicators of renal function or disease. *Journal of Veterinary Internal Medicine* 2013;27:31–38.
64. Tefft KM, Shaw DH, Ihle SL et al. Association between excess body weight and urine protein concentration in healthy dogs. *Veterinary Clinical Pathology* 2014;43:255–260.
65. Cameron ME, Casey RA, Bradshaw JW et al. A study of environmental and behavioural factors that may be associated with feline idiopathic cystitis. *Journal of Small Animal Practice* 2004;45:144–147.
66. Defauw PAM, Van de Maele I, Duchateau L et al. Risk factors and clinical presentation of cats with feline idiopathic cystitis. *Journal of Feline Medicine and Surgery* 2011;13:967–975.
67. Lund HS, Krontveit RI, Halvorsen I et al. Evaluation of urinalyses from untreated adult cats with lower urinary tract disease and healthy control cats: Predictive abilities and clinical relevance. *Journal of Feline Medicine and Surgery* 2013;15:1086–1097.
68. Jones BR, Sanson RL, Morris RS. Elucidating the risk factors of feline lower urinary tract disease. *New Zealand Veterinary Journal* 1997;45:100–108.
69. Lekcharoensuk C, Osborne CA, Lulich JP. Epidemiologic study of risk factors for lower urinary tract diseases in cats. *Journal of the American Veterinary Medical Association* 2001;218:1429–1435.
70. Segev G, Livne H, Ranen E et al. Urethral obstruction in cats: Predisposing factors, clinical, clinicopathological characteristics and prognosis. *Journal of Feline Medicine and Surgery* 2011;13:101–108.
71. Kennedy SM, Lulich JP, Ritt MG et al. Comparison of body condition score and urinalysis variables between dogs with and without calcium oxalate uroliths. *Journal of the American Veterinary Medical Association* 2016;249:1274–1280.
72. Subak LL, Richter HE, Hunskaar S. Obesity and urinary incontinence: Epidemiology and clinical research update. *Journal of Urology* 2009;182:S2–S7.
73. Wynn SG, Witzel AL, Bartges JW et al. Prevalence of asymptomatic urinary tract infections in morbidly obese dogs. *PeerJ* 2016;4:e1711.
74. Kapoor M, Martel-Pelletier J, Lajeunesse D et al. Role of proinflammatory cytokines in the pathophysiology of osteoarthritis. *Nature Reviews: Rheumatology* 2011;7:33–42.
75. Backus R, Wara A. Development of obesity: Mechanisms and physiology. *The Veterinary Clinics of North America: Small Animal Practice* 2016;46:773–784.
76. Frank L, Mann S, Levine CB et al. Increasing body condition score is positively associated interleukin-6 and monocyte chemoattractant protein-1 in Labrador retrievers. *Veterinary Immunology and Immunopathology* 2015;167:104–109.
77. Collins KH, Paul HA, Reimer RA et al. Relationship between inflammation, the gut microbiota, and metabolic osteoarthritis development: Studies in a rat model. *Osteoarthritis and Cartilage* 2015;23:1989–1998.

78. Edney AT, Smith PM. Study of obesity in dogs visiting veterinary practices in the United Kingdom. *The Veterinary Record* 1986;118:391–396.
79. Huck JL, Biery DN, Lawler DF et al. A longitudinal study of the influence of lifetime food restriction on development of osteoarthritis in the canine elbow. *Veterinary Surgery* 2009;38:192–198.
80. Smith GK, Paster ER, Powers MY et al. Lifelong diet restriction and radiographic evidence of osteoarthritis of the hip joint in dogs. *Journal of the American Veterinary Medical Association* 2006;229:690–693.
81. Smith GK, Mayhew PD, Kapatkin AS et al. Evaluation of risk factors for degenerative joint disease associated with hip dysplasia in German Shepherd Dogs, Golden Retrievers, Labrador Retrievers, and Rottweilers. *Journal of the American Veterinary Medical Association* 2001;219:1719–1724.
82. Grierson J, Asher L, Grainger K. An investigation into risk factors for bilateral canine cruciate ligament rupture. *Veterinary and Comparative Orthopaedics and Traumatology* 2011;24:192–196.
83. Adams P, Bolus R, Middleton S et al. Influence of signalment on developing cranial cruciate rupture in dogs in the UK. *Journal of Small Animal Practice* 2011;52:347–352.
84. Impellizeri JA, Tetrick MA, Muir P. Effect of weight reduction on clinical signs of lameness in dogs with hip osteoarthritis. *Journal of the American Veterinary Medical Association* 2000;216:1089–1091.
85. Mlacnik E, Bockstahler BA, Müller M et al. Effects of caloric restriction and a moderate or intense physiotherapy program for treatment of lameness in overweight dogs with osteoarthritis. *Journal of the American Veterinary Medical Association* 2006;229:1756–1760.
86. Marshall WG, Hazewinkel HAW, Mullen D et al. The effect of weight loss on lameness in obese dogs with osteoarthritis. *Veterinary Research Communications* 2010;34:241–253.
87. Bennett D, Zainal Ariffin SM, Johnston P. Osteoarthritis in the cat: 1. How common is it and how easy to recognise? *Journal of Feline Medicine and Surgery* 2012;14:65–75.
88. Keller GG, Reed AL, Lattimer JC et al. Hip dysplasia: A feline population study. *Veterinary Radiology & Ultrasound* 1999;40:460–464.
89. Jensen VF, Ersbøll AK. Mechanical factors affecting the occurrence of intervertebral disc calcification in the dachshund – A population study. *Journal of Veterinary Medicine. A, Physiology, Pathology, Clinical Medicine* 2000;47:283–296.
90. Levine JM, Levine GJ, Kerwin SC et al. Association between various physical factors and acute thoracolumbar intervertebral disk extrusion or protrusion in Dachshunds. *Journal of the American Veterinary Medical Association* 2006;229:370–375.
91. Canal S, Contiero B, Balducci F et al. Risk factors for diskospondylitis in dogs after spinal decompression surgery for intervertebral disk herniation. *Journal of the American Veterinary Medical Association* 2016;248:1383–1390.
92. Anfinsen KP, Grotmol T, Bruland OS et al. Primary bone cancer in Leonbergers may be associated with a higher bodyweight during adolescence. *Preventive Veterinary Medicine* 2015;119:48–53.

93. Sherwood JM, Haynes AM, Klocke E et al. Occurrence and clinicopathologic features of splenic neoplasia based on body weight: 325 dogs (2003–2013). *Journal of the American Animal Hospital Association* 2016;52:220–226.
94. Lim HY, Im KS, Kim NH et al. Obesity, expression of adipocytokines, and macrophage infiltration in canine mammary tumors. *Veterinary Journal* 2015;203:326–331.
95. Lim H-Y, Im K-S, Kim N-H et al. Effects of obesity and obesity-related molecules on canine mammary gland tumors. *Veterinary Pathology* 2015;52:1045–1051.
96. Weeth LP, Fascetti AJ, Kass PH et al. Prevalence of obese dogs in a population of dogs with cancer. *American Journal of Veterinary Research* 2007;68:389–398.
97. Michel KE, Sorenmo K, Shofer FS. Evaluation of body condition and weight loss in dogs presented to a veterinary oncology service. *Journal of Veterinary Internal Medicine* 2004;18:692–695.
98. Romano FR, Heinze CR, Barber LG et al. Association between body condition score and cancer prognosis in dogs with lymphoma and osteosarcoma. *Journal of Veterinary Internal Medicine* 2016;30:1179–1186.
99. Kadry B, Press CD, Alosh H et al. Obesity increases operating room times in patients undergoing primary hip arthroplasty: A retrospective cohort analysis. *PeerJ* 2014;2:e530.
100. Kabon B, Nagele A, Reddy D et al. Obesity decreases perioperative tissue oxygenation. *Anesthesiology* 2004;100:274–280.
101. Hedenstierna G, Santesson J. Breathing mechanics, dead space and gas exchange in the extremely obese, breathing spontaneously and during anaesthesia with intermittent positive pressure ventilation. *Acta Anaesthesiologica Scandinavica* 1976;20:248–254.
102. Salome CM, King GG, Berend N. Physiology of obesity and effects on lung function. *Journal of Applied Physiology* 2010;108:206–211.
103. Jones RL, Nzekwu MM. The effects of body mass index on lung volumes. *Chest* 2006;130:827–833.
104. Bach JF, Rozanski EA, Bedenice D et al. Association of expiratory airway dysfunction with marked obesity in healthy adult dogs. *American Journal of Veterinary Research* 2007;68:670–675.
105. Manens J, Ricci R, Damoiseaux C et al. Effect of body weight loss on cardiopulmonary function assessed by 6-minute walk test and arterial blood gas analysis in obese dogs. *Journal of Veterinary Internal Medicine* 2014;28:371–378.
106. García-Guasch L, Caro-Vadillo A, Manubens-Grau J et al. Pulmonary function in obese vs non-obese cats. *Journal of Feline Medicine and Surgery* 2015;17:494–499.
107. Ferretti A, Giampiccolo P, Cavalli A et al. Expiratory flow limitation and orthopnea in massively obese subjects. *Chest* 2001;119:1401–1408.
108. Manens J, Bolognin M, Bernaerts F et al. Effects of obesity on lung function and airway reactivity in healthy dogs. *Veterinary Journal* 2012;193:217–221.

109. Mosing M, German AJ, Holden SL et al. Oxygenation and ventilation characteristics in obese sedated dogs before and after weight loss: A clinical trial. *Veterinary Journal* 2013;198:367–371.
110. Iliescu R, Tudorancea I, Irwin ED et al. Chronic baroreflex activation restores spontaneous baroreflex control and variability of heart rate in obesity-induced hypertension. *American Journal of Physiology: Heart and Circulatory Physiology* 2013;305:H1080–H1088.
111. Mehlman E, Bright JM, Jeckel K et al. Echocardiographic evidence of left ventricular hypertrophy in obese dogs. *Journal of Veterinary Internal Medicine* 2013;27:62–68.
112. Marshall W, Bockstahler B, Hulse D et al. A review of osteoarthritis and obesity: Current understanding of the relationship and benefit of obesity treatment and prevention in the dog. *Veterinary and Comparative Orthopaedics and Traumatology* 2009;22:339–345.
113. Summers LK, Samra JS, Humphreys SM et al. Subcutaneous abdominal adipose tissue blood flow: Variation within and between subjects and relationship to obesity. *Clinical Science (London)* 1996;91:679–683.
114. Steagall PV, Pelligand L, Giordano T et al. Pharmacokinetic and pharmacodynamic modelling of intravenous, intramuscular and subcutaneous buprenorphine in conscious cats. *Veterinary Anaesthesia and Analgesia* 2013;40:83–95.
115. Ingrande J, Lemmens HJ. Dose adjustment of anaesthetics in the morbidly obese. *British Journal of Anaesthesia* 2010;105:i16–i23.
116. Hanley MJ, Abernethy DR, Greenblatt DJ. Effect of obesity on the pharmacokinetics of drugs in humans. *Clinical Pharmacokinetics* 2010;49:71–87.
117. Cheymol G. Effects of obesity on pharmacokinetics implications for drug therapy. *Clinical Pharmacokinetics* 2000;39:215–231.
118. Boveri S, Brearley JC, Dugdale AH. The effect of body condition on propofol requirement in dogs. *Veterinary Anaesthesia and Analgesia* 2013;40:449–454.
119. Brill MJ, Diepstraten J, van Rongen A et al. Impact of obesity on drug metabolism and elimination in adults and children. *Clinical Pharmacokinetics* 2012;51:277–304.

4 通过评价体成分诊断宠物肥胖症

4.1 引言

犬猫肥胖症的主要因素,似乎是未能及时发现超重或肥胖状态所导致的营养失调。宠物主人往往不能正确地评价其宠物的肥胖程度,尤其常常会低估超重和肥胖宠物的体况评分[1-5]。因此,兽医对宠物超重或肥胖的鉴别,对于适时预防和治疗干预起到重要作用。

然而,在大型转诊中心和较小的私人兽医诊所中,通常都不会记录患病宠物的体况评分,即使做了相关记录,也常常缺少针对低于或高于正常体况评分时的干预措施[6-9]。为此,2011年世界小动物兽医协会(World Small Animal Veterinary Association,WSAVA)将营养评价作为宠物的第5个生命体征[10]。对于所有到兽医处就诊的宠物,临床医生应该像测量体温、脉搏、呼吸频率和疼痛评价一样,将营养评价(含体况评分)纳入必检项目。此外,从2014年美国动物医院协会(American Animal Hospital Association,AAHA)的体重管理指南中,开始建议兽医在犬猫的生长期、绝育后,以及随后每年的体检期间监测体重变化趋势,并在体况评分高于理想值时,尽早主动解决体重增加问题[11]。当前有几种方法可用于评估宠物的体成分,有些较为客观,但价格昂贵且需要专用设备;有些则较便宜,但更易于在临床诊治环境中使用。若要正确地使用这些不同的工具和方法,必须了解它们的优势和局限性,并了解每种方法的准确度(提供反映真实值或绝对值的正确测量结果的能力)和精密度(提供可靠且一致的测量结果的能力)。下文将描述和讨论在犬猫活体状态和死后解剖情况下,体成分最常用和最佳的评价方法。

4.2 体成分的胴体分析

胴体分析是一种测量体成分的化学分析方法,被认为是评价各种动物体成分的黄金标准[12]。该方法的步骤为:先解剖所有组织,然后将组织进行研磨和均质化(匀浆);通过烘箱烘干或冷冻干燥(冻干)已知重量的组织,并测量干燥前后的重量,根据差值计算含水量;再使用有机溶剂测定并计算脂肪含量,通过燃烧样品中全部有机物确定矿物质含量[12,13]。对于可以分析整个动物胴体,且能在安乐死后立即进行试验的小型动物而言,这是一种非常准确的方法。但是,这种方法容易出现 2 个问题,一是如果胴体在分析之前被冷冻过,水分蒸发可能会影响测定结果;二是若研究对象是较大的动物,则只能分析其局部小样本,可能无法代表整个胴体的体成分组成[13,14]。另外,在处理完整的胴体时,胃肠道(gastrointestinal,GI)内容物也可能会影响测定结果。然而,这 2 个问题可以通过以下方法解决:在冷冻前后分别对动物称重计算失水量,以及在动物分析前禁食,在动物处死后切除内脏或排空肠道[13]。因为胴体分析方法只能在动物安乐死之后使用,所以主要在科学研究和验证其他方法时使用,不能用于兽医诊治。

其他大多数评价体成分的技术都是基于间接方法,原理是利用至少 2 种化学成分迥异的物质——通常包括脂肪和非脂肪物质(或称去脂体重,fat-free mass,FFM),建立的体成分模型[14]。然而,如表 4.1 所示,不同的测定方法采用了不同的体成分指标[15]。

表 4.1　体成分术语、含义及犬猫体成分常用测量方法

术语	解释	测量方法
体脂量	体内甘油三酯脂肪量	胴体分析 双能 X 射线吸收法 (生物电阻抗分析) 活体测量 体况评分
脂肪组织量	脂肪(约 83%)以及支持细胞和细胞外结构(约 2% 的蛋白质和约 15% 的水组成)	计算机断层扫描 核磁共振成像
瘦体重	非脂肪组织体重(包括骨矿物质)	D_2O 重水法 计算机断层扫描 核磁共振成像
非脂肪组织/ 去脂体重	瘦体重以及脂肪组织的脱脂成分	胴体分析 生物电阻抗分析 (双能 X 射线吸收法)
体细胞质量	身体的细胞成分	D_2O 重水法
细胞外固体	全身骨矿物质(骨骼约占细胞外固体的 85%)和筋膜、软骨等(约占细胞外固体的 15%)	胴体分析 双能 X 射线吸收法

资料来源:Jensen MD. *Journal of the American Dietetic Association*,1992;92(4):454-460.
注:测量方法写在括号内时,表示根据使用该方法获得的其他测量值计算得到。

4.3 影像学方法评估体成分

4.3.1 双能 X 射线吸收仪

双能 X 射线吸收仪(dual energy x-ray absorptiometry,DEXA)最初开发并应用于评估人类骨矿物质含量,目前已发展成为用于体成分评价的有效工具[16,17]。在兽医研究中,已证明双能 X 射线吸收仪非常适用于评估骨矿物质含量和体成分,可测定骨量、脂肪量和瘦软组织量[18](图 4.1)。在使用双能 X 射线吸收仪时,术语"瘦体重"和"瘦软组织量"通常不做区分。但是,根据瘦体重的定义,其包括用 DEXA 单独测定的骨矿物质[19]。而由于缺乏对机体不同腔室的明确定义,这可能会导致由于特定腔室的重叠或遗漏而造成数据报告的错误。如图 4.1 所示,在已明确提出的各个术语的定义中[19-21],建议将"瘦体重"与"去脂体重"列为同义词。

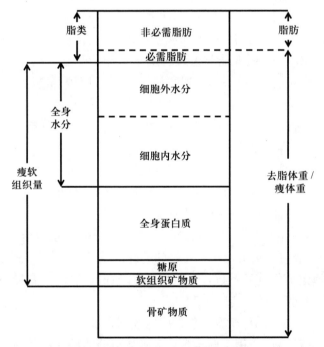

图 4.1 体成分评价方法和术语。关于瘦体重的定义存在差异。
本图中,瘦体重被定义为等同于去脂体重

(改编自 Heymsfield S B,et al. *Annual Review of Nutrition* 1997;17:527-558)

双能 X 射线吸收仪的物理原理是 2 束 X 射线穿过机体后,测定它们在低光子和高光子能量水平下的衰减。通过使用不同的算法,双能 X 射线吸收仪可以对每个像素的不同组织类型进行量化,从而估计出总质量、骨矿物质含量、脂肪和瘦软组织量[15,17,22]。DEXA在扫描期间的辐射积存量很低[17]。Pietrobelli等撰写的综述详细介绍了双能X射线吸收仪测定体成分的物理概念[20]。

生产双能X射线吸收仪测定系统的制造商数量不多,虽然吸收仪都基于相同的物理原理,但它们在高能和低能X射线束的产生、X射线探测器、几何成像、校准方法和软件算法方面有所不同[22]。这些系统提供了包括针对小动物、儿童和成年人不同的分析软件包。不同设备之间不可避免的存在差异,因此,在比较骨矿物质含量和体成分评价结果时,应谨慎判断[23-25]。

双能X射线吸收仪已对多个物种的胴体化学分析进行过比较验证[12,26-34]。使用DEXA估测的总质量与使用体重秤的测量值,显示出一贯的高度相关性[12,28-33,35]。对犬体成分的研究表明,DEXA和化学分析两者得出的结果,体脂率和非脂肪组织百分比之间没有显著差异[26]。然而,在测量骨矿物质含量时,双能X射线吸收仪和灰分百分含量测定法之间存在显著差异[26]。在Speakman等的研究中,犬和猫2种动物之间的总体重($r^2 = 1.00$)、去脂体重($r^2 = 0.998$)和脂肪量($r^2 = 0.964$)的评估值存在强相关性。犬与猫之间的非脂肪组织和脂肪组织的平均误差较小,分别为2.64%和2.04%。然而,非脂肪组织的差异范围从低估 −2.64% 到高估 13.22%,脂肪量的差异范围从低估 −20.7% 到高估 31.5%,原因可能与犬和猫之间肌肉的水合状态不同有关[12]。有学者对犬猫体重范围内的儿童(译者注:这里原文为"pig"——猪,但根据上下文和参考文献,推测应为儿童)和仔猪开展了研究,尽管观察到绝对脂肪量存在显著差异,但双能X射线吸收仪测量值与化学分析值之间显示出了高度相关性[27-29,31-33]。这些差异被认为是由不同仪器制造商和分析软件之间的差异造成的[24,32],也可能是因为用于化学分析的胴体存在均质化不完全的情况[27]。更新的仪器、更完善的软件系统以及具有特异性的校准系数,可进一步提高该方法的准确性[29,36,37]。对猫、犬和猪的验证研究发现,双能X射线吸收仪技术在测量体成分腔室方面显示出了良好的精确性[26-28,32,38-40]。然而,在已报道的研究中,脂肪量测定值的变异系数(coefficient of variation,CV)一直最差,例如在一项针对猫的研究中,变异系数接近6%[38]。

在犬猫体成分分析中,双能X射线吸收仪通常被认为是仅次于胴体分析的第二好的方法。其已被广泛用于确定各种犬种体成分的种内和种间差异[12,26],用以估计犬和猫体成分和体重变化[40-44],以及用于验证评价体成分的其他方法,如生物电阻抗、体脂指数系统和体况评分[45-52]。

4.3.2 计算机断层扫描

与DEXA一样,计算机断层扫描(computer tomography,CT)也是基于X射线技术,但其辐射负担较大。在 CT 扫描过程中,可以从多个角度获得 X 射线衰减的测量值。随后的 CT 重建将原始采集的数据转换为一系列 CT 图像,每个图像由若干体积单元(volume elements)或体素(voxels)构建。每个体素中的平均线性衰减系数(CT 值)使用霍恩斯菲尔德单位(Hounsfield unit,HU)表示,其中,0 HU 对应为水的线性衰减系数,-1 000 HU 对应为空气的线性衰减系数。由于物理密度是衰减的主要决定因素之一,而各个组织,例如脂肪组织和软组织等,具有各自独特的组织密度,因此可以通过不同的 HU 数值进行区分。使用精细的 CT 图像和体素评价方法,不仅可以确定瘦组织和脂肪组织,还可以确定组织分布,例如,估算内脏和皮下脂肪室的分布情况[53]。

尽管在兽医临床治疗中,使用 CT 扫描仪的数量正在增加,但应用于伴侣动物体成分的研究仍数量有限,其有效性验证研究也非常有限。Ishioka 等根据重水(D_2O)稀释技术验证了 7 只比格犬的 CT 扫描结果,提出了犬特定脂肪组织的 CT 值范围为 -135~-105 HU,并提出在第三腰椎(L3;$r^2 = 0.96$)处水平作为采样点对评估体脂含量的效果较好[54]。猫脂肪组织的 CT 值范围已确定为 -156~-106 HU,其中 L3(第三腰椎)和 L5(第五腰椎)为整个腹部位置的最佳采样点[55]。为了克服 Ishioka 等和 Lee 等提出的 HU 范围组织特异性的限制[54,55],有学者使用 CT 值研究了体素的频率分布,由此可以建立猫的脂肪组织百分比与 DEXA 获得的体脂百分比的相关性[56]。CT 也被用于评估内脏和皮下脂肪量与体重增长的关系,以及与绝育和代谢性疾病中不同生物标志物相关的体成分变化[54,57,58]。

4.3.3 核磁共振成像

与 DEXA 和 CT 相比,核磁共振成像(magnetic resonance imaging,MRI)不使用诸如 X 射线的电离辐射。而是基于产生的磁场和生物组织中大量的质子(即氢核)之间的相互作用[53]。

MRI 可获得断层图像。由于不同类型的组织(例如脂肪组织、脑实质等)具有不同的局部核磁共振特性,因此 MRI 可提供对解剖差异具有高度敏感性的高对比度图像[53]。

伴侣动物的 MRI 与体成分相关的出版物数量很有限。其中,有结合半自动图像分析,使用 MRI 对猫肥胖的评价[59],以及将全身脂肪-水 MRI(fat-water MRI,FWMRI)成功地应用于犬增重前后的评价[60]。FWMRI 估算的总体重与体重秤的测量结果一致性很高($r^2 = 0.867$),并且表明随着体重的增加,内脏脂肪、皮下脂肪和总脂肪重显著增加。FWMRI 是一种可精确测量总脂肪组织、内脏组织、瘦组织和骨密度的方法,其变异系数(CV)在 0.2%~3.1%[60]。

使用一种称为定量核磁共振(quantitative magnetic resonance,QMR)的新技术,

也可对犬和猫的体成分进行 D_2O 和 DEXA 检测。结果表明,定量核磁共振是一种很有应用前景的体成分测量工具,但与 D_2O 相比,QMR 测得犬的总体重、瘦体重和脂肪量明显偏低[61]。而在对猫的研究中,QMR 与 D_2O 结果相比,仅发现脂肪量被显著低估[62]。除了可避免电离辐射外,QMR 还可以在未镇静的动物上进行。通过使用适当的校正方程,QMR 有望成为能够准确评价体成分的工具[61,62]。

影像学方法是一种客观且可重复的无创评价方法。但诸如 DEXA、CT 和 MRI 之类的专用设备,仅限于大型兽医中心和大学的宠物医院使用。此外,通常还需要对动物进行全身麻醉或镇静处理,而且某些方法还会给动物和工作人员带来辐射负担(图 4.2)。

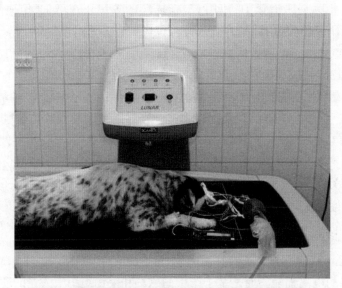

图 4.2 麻醉犬腹卧位正在接受双能 X 射线吸收仪扫描。
在扫描过程中,大扫描臂从患宠身体上方移动经过。
扫描台上有脉搏血氧测定设备(照片由 A.D. Vitger 提供)

4.4 非影像学方法评估体成分

4.4.1 重水稀释技术

重水(D_2O)稀释技术的理论基础是机体水分主要与非脂肪组织相关,因此可以通过测量总体水(全身所含水分的总量,total body water)来间接测量去脂体重[13]。D_2O 是一种稳定无毒的同位素指示剂,可以与水自由交换。在静脉内注射已知量的 D_2O 之后,在一定的时间内,D_2O 将均匀地分布在身体各处。当 D_2O 和水达到平衡时,可以通过测量血液样本中同位素指示剂的稀释程度来评估体内总水量[13,63]。基于非脂肪组织中包含73.2%水分的假设,可以使用以下公式计算体脂率(body fat percentage,BF):体脂率 BF = 100 – 总体水的百分数 /0.732[13,46,63]。但是有人提出质疑,认为经肺呼吸和尿液损失的同位素是不可避免的,并可能会影响测定结果。然而,在一项对犬的研究中,测定出尿中同位素的流失可忽略不计[13,63]。与胴体分析相比,D_2O 高估了21.8%的总体水含量,评价值相对较高,并且由于非参数数据的存在,一些方程无法得到验证。但是,基于 D_2O 方法所提出的方程,机体含水率($r^2 = 0.96$)、体脂率($r^2 = 0.96$)、含氮率($r^2 = 0.77$)和含灰分率($r^2 = 0.64$)呈现良好的相关性[13]。对犬的研究表明,与 D_2O 相比,DEXA 对体脂高估了13%~15.8%。尽管缺乏绝对一致性,但这2种方法显示出了良好的相关性($r^2 = 0.78$)[35,46]。虽然有人担心 BF 可能存在着系统性低估,但大多数研究表明,犬和猫的 D_2O 法、胴体分析法和 DEXA 法之间有良好的相对一致性[35,46,61-63]。但是,D_2O 法中同位素平衡所需的时间会影响测定结果,因此应将其标准化[13,64]。

重水稀释技术虽然是一种安全无创的方法,可定量测量犬猫的体成分。但是该方法费用昂贵,需要住院至同位素达到平衡,因此仍主要在科学研究中使用。

4.4.2 生物电阻抗

生物电阻抗(bioampedance,BIA)的原理是测量肌肉和脂肪组织中水分对电信号的电阻。较高的肌肉和较低的脂肪含量会导致较高的体内水分含量,这有利于电流通过身体。如果脂肪较多,则电阻会增加,电流会降低[65,66]。该方法被称为是一种安全、无创、易于使用的"床边测试"方法[65]。然而,该方法依赖于这样一个假设,即受试者具有相似的身体几何结构以及恒定的非脂肪组织组成。因此,动物机体的水合状态,以及与身体形态改变(如肥胖)相关的变化,均可能对测定结果产生重大影响[65,66]。为了估算去脂体重,可以使用单频和多频阻抗,并可开发出针对特定群体的经验回归模型[65]。在体重指数 BMI 小于34的健康人群中,BIA 被认为是一种合理可靠的体成分筛选测试方法,但如果 BMI 较高,则应谨慎使用该方法[67]。对于犬猫的相关研究很少,结果也有争议。在一项对22只猫进行的研究中,通过与化学分析的比较,发现单频 BIA 结合特定的动物测量方法能够可靠的预测去脂体重[68]。在另两项研究中,将猫全身和细胞外水分的多频 BIA 估计值,与 D_2O 和溴化物的稀释估计值进行了比较[69,70]。两项研究均表明该方法是有效的,但结果取决于动物体态的情况($r^2 = 0.17$~0.71)[69,70]。有2项对犬的研究对单频 BIA 和 DEXA 估测的体脂率进行了比较[45,71],发现尽管生物电阻抗法具有较好的精确度,并与 DEXA 法具有良好的相关性($r^2 = 0.44$~0.98),但各犬只之间的精确度存在差异,并且2种方法之间的一致性差,从而得出结论:在未经进一步验证的情况下,该方法不能应用于临床[45,71]。

4.5 临床环境下的体成分评估

若要在兽医诊所启动和监测减肥干预措施,首先必须找到可靠的方法来评估患病宠物的肥胖程度。在实际操作中,很难使用 D_2O、MRI、DEXA 或 CT 来评估体成分。这些方法耗时且昂贵,并且患宠通常难以接受麻醉或镇静。理想的用于估算体成分的方法应该简单、准确和精确,并且重要的是这些方法要快速实用,可在初次诊治中应用。

4.5.1 体重

体重可以用适合动物大小的体重秤客观地测量。理想的情况是,宠物每次就诊应使用相同的体重秤,因为体重秤之间的差异可能会影响测量结果。体重很容易测量,可客观给出可重复的数值结果,是检测单只患宠体重细微变化,以及评估正在进行的体重管理方案的优秀工具。因此,在患病宠物每次就诊时兽医都应记录其体重。

然而,尽管体重能够准确地体现出动物的总重量,但其在评估体成分方面的能力有限[68,72,73]。理想体重取决于宠物的品种、公母、运动水平和身高,即使在同一品种内,个体之间也存在差异。

4.5.2 形态测量

4.5.2.1 体重指数

人类的体重指数可根据人的身高和体重来估算体成分,受之启发,由此开发出了犬猫的体重指数。这种方法可提供体成分的数值评估,而不是体况评分中更为严格的分类结果,因此在没有更先进的研究方法可用时,此方法很有吸引力。

4.5.2.1.1 犬的体重指数

为了评估犬的 BMI 和特定性别的体脂率(BF),有学者提出了通过形态测量建立的方程[74]。作者认为,当犬站直并向前看的情况下,容易进行形态测量[46]。BF 计算结果显示出良好的相关性(r^2 = 0.9~0.92),并且与 DEXA 结果一致,而 BMI 计算值尽管与 DEXA 有显著相关性(r^2 = 0.54),但对于个体值的一致性较差[46,71]。在对不同基因品种的犬进行的一项研究中,体脂率 BF 方程与 DEXA 结果之间的相关性较低(r^2 = 0.49~0.58),而且个体测量结果的一致性不可接受[71]。特定品种的计算方法更适合某些品种,将形态测量方法与 BIA 结合使用有一定的前景,但是需要对不同基因犬种进行更多研究,以确定全部或特定品种的方程是否能可靠地估算犬的体脂率。

4.5.2.1.2 猫的体重指数

人们还提出了猫体重指数的概念(feline body mass index,FBMI)[75,76]。据 Hawthorne 和 Butterwick 研究,与体况评分(r^2 = 0.73)相比,此方法预测的 BF 与 DEXA 测量的相关性(r^2 = 0.85)更高。在最近的一项研究中,人们发现 FBMI 和 DEXA 估测的 BF 之间具有很好的相关性(r^2= 0.85)[77],但是,这 2 种方法对于个体 BF 的差异均超过 10%,并且身高测量值很容易由于猫的配合程度产生几厘米的差异。根据 FBMI 测定方法的要求,猫应处于站立状态,腿垂直于桌面,头处于直立位置[75]。根据我们的经验,这种姿势很难实现,这使得 FBMI 在猫临床诊疗实际操作中很少使用。

4.5.2.2 新形态方程

田纳西大学与希尔斯宠物营养公司(Hill's Pet Nutrition,位于美国堪萨斯州托皮卡市)合作提出了新形态方程[51,52]。对于犬的新形态方程,估算瘦体重(而不是之前在 DEXA 讨论过的图 4.1 中所述的瘦软组织质量)的最佳拟合方程的 r^2(相关系数)可达 0.98,估算脂肪量和 BF 的 r^2 可达 0.98 和 0.82[52]。对于猫,估算瘦体重和脂肪量的 r^2 分别为 0.85 和 0.98[52]。研究人员发现 2 个系统的测试动物之间(差异 1%~19.5%)和测试动物自身(差异 <2%)的差异性都相对较低,并且每个动物的所有测量都预期可以在 5 min 内完成。但是,有些犬猫需要镇静才能完成所有测量[51,52]。仅有少数几个品种的猫适合此方法,而且犬的大小和体型范围广泛,这些都对开发评估体成分的标准化方程提出了重大挑战。因此,这些形态测定方程仍有待在单独的研究群体中进行验证,以确定其是否可适用于全球范围内的宠物群体。

4.5.3 形态学估算

4.5.3.1 体况评分

犬猫的体况评分系统是一种带有主观性的半定量方法,是通过视觉和触觉特征来估计犬猫的肥胖程度。通过视诊和触诊,可对宠物的脂肪沉积情况进行评估,然后从已预先定义的分值等级范围推断出体况评分。犬猫通常通过评估胸腔的脂肪覆盖程度、有无腰部形态以及腹部脂肪垫的大小来确定体况评分。人们已经提出了几种体况评分系统[47,48,78-81]。当使用半分制时,5 分制系统类似于 9 分制系统[11],这些提出的系统尽管都没有得到广泛验证,但似乎 5 分制和 9 分制系统的使用最广泛,其中 1 分代表体况消瘦,3 或 5 分表示体况理想,而 5 或 9 分为严重肥胖[1,45,51,52,71-73,82-88]。根据 9 分制等级设定的体况评分与 DEXA 测量的 BF 相比,不论流浪猫和家养猫(r^2 = 0.64~0.83),还是流浪犬和家养犬(r^2 = 0.58~0.92),都有良好的相关性[45-49,73,78]。根据该系统在犬猫中的初步应用,可以确定理想的 BCS 为 5/9 时,犬的脂肪量为(19 ± 8)%,而在雄性猫中,BCS 为 5/9 时,其脂肪量为(21.8±1.7)%,并且观察到犬猫 2 个物种的 BCS 每增加 1 分,其脂肪量都增加 5%[47,78,89]。尽管有相似的相关性,然而每个 BCS 分值单位的 BF,在不同种群之间可能有所不同。在犬中,Mawby 等发现理想的 BCS 为 5/9,其脂肪量为(11±2)%,并且 BCS 每增加 1 分,脂肪量就会增加 8.7%[46]。本书作者的一项小组研究发现,家养且已绝育、BCS 为 5/9 的雄性和雌性猫的 BF 中位数为 31.8%(范围为 23.6%~38.4%),BCS 每增加 1 分,其 BF 相应增加 5%~6%[49],而在另一项对流浪猫的研究中,BCS 为 3/5 时,其 BF 为(11.7±4.5)%,每升高 1/2 单位 BCS 对应 BF 增加 7%[73]。研究人员使用不同的 DEXA 硬件和软件系统,或测定体况评分时人为的技术差异,可能是部分导致不同报告中 BF 水平差异的原因。然而,在先前的体况评分研究中,观察者自身的变异系数(8.1%)和观察者之间的变异系数(11.6%~15%)都相对较低[47,48]。2 项研究之间的差异可能是:与户外活动的流浪犬猫或家养犬猫相比,室内长期缺乏运动宠物的肌肉量有所减少[90],而且各个独立研究中群体选择的差异也可能会影响结果[71]。在理想的 BCS 为 5/9 的情况下,灵缇犬的 BF(7.2%)显著低于哈士奇犬(31%),并且也往往比罗威纳犬(32%)更低,而与贵宾犬(18%)和腊肠犬(15%)相比差异不显著[71]。此外,根据品种的不同,BCS 每增加 1 分,BF 平均增加 1.5%~6.5%。这些发现突显了开发特定品种的体况评分图表的重要性。皇家宠物食品公司(Royal Canin)提出了特定品种犬猫体型的图表,包括所有 9 个等级 BCS 的图示[91],但据作者所知,这些图表尚未得到验证。

应该注意的是,所有的 BCS 系统都已经在群体中进行了开发和验证,但 BCS<4/9 的宠物很少甚至没有,BCS 高于 8/9 的宠物也相对较少,同样的情况也包括脂肪量高达 45%、BCS 为 9/9 或 5/5 的宠物[46,47,49,71,73,78,89]。

最近,有学者评估了 5 分制 BCS 系统与犬猫体脂指数(body fat index,BFI)的关系,发现 BCS 系统效果不如 BFI 和形态测量学系统[51,52]。根据定义,在使用 BCS 系统的情况下,所有认定 BCS 为 5/5 的动物都被评估为有 40% 的体脂率。因此,这项研究中 BCS 的不佳表现很可能反映了这样一个事实:在选定的群体中,65% 的犬和 71% 的猫的 BF>40%,受 BCS 所使用的定义所限,所有这些犬猫的 BF 自然会被低估。

体况评分的准确性很大程度上取决于研究人员的培训情况。未经培训的评分人或宠物主人通常会低估其宠物的 BCS,与之相比,兽医或经过培训的评分人所提供的 BCS 会更准确[2,4,5,73]。在一项研究中,宠物主人使用 5 分制 BCS 系统来评估宠物的体况,结果并未提高与训练有素的研究人员测定结果的一致性[3]。因此需要进一步研究,以评估如何最好地引导宠物主人识别和监测宠物的体重变化。

BCS 系统易于在临床诊疗中使用,其具有说服力和可重复性,并可作为一个有价值的工具来评估宠物体重是否正常、不足或超重。此外,有研究者在规划和监测宠物个体减肥计划时增加了体重测量[92]。需要注意的是，BCS 为 9 的犬猫可能会呈现出很大的体脂率范围，使得准确评估这些患宠的理想体重并非易事。

4.5.3.2 犬猫体脂指数

为了解决肥胖宠物 BF 的评估问题，有研究者已开发出 2 套补充系统，分别针对犬猫进行了验证[51,52]。这些系统仍然是基于对动物身体指定区域的视诊和触诊。其涵盖了从轻度超重（BFI 30，BF 25%~35%）到极度肥胖（BFI 60，BF 56%~65%）。猫或犬的肥胖程度可以根据 6 幅插图和书面描述进行分类，然后根据当前体重和 BFI 类别，从图表中估算出理想体重[51,52]。研究人员发现，有 53% 的犬使用 BFI 系统估测的 BF 与 DEXA 测定值的误差在 10% 以内，有 91% 的犬估测的 BF 与 DEXA 测定值的误差在 20% 以内。对猫的 BF 而言，其 BFI 系统估测值与 DEXA 测定值误差在 10% 和 20% 以内的比例分别为 56% 和 90%。但是这 2 种方法之间的相关性尚未见报道。被纳入的犬和猫只有少数是体重过轻或过瘦的（占 6% 的 5 只犬的 BF<25%，占 8% 的 6 只猫的 BF<30%），因此，当确定超重或肥胖的犬猫时，这 2 种 BFI 系统的应用将是对 BCS 的补充。

4.6 全科诊疗的临床建议

体况评分系统是体重监测的补充，可在更易于实施预防措施的阶段识别出有肥胖风险的宠物。对兽医实践而言，应重视在所有诊治过程中采取称重和体成分评估，并对偏离理想体重的情况采取措施。尤其是在宠物绝育后，应建议宠物主人减少食物供应，并定期监测宠物的体重。肌肉状况评分系统已经被提出，其将是 BCS 系统的理想补充，然而，它们在肥胖症治疗中的应用还需要进一步开发和验证[10,50]。

或许，最终将生物电阻抗（BIA）与形态测量学评价方法结合使用，可以提供更准确的方法来估算不同品种和居住环境下犬猫的去脂体重，但这仍有待进一步研究。对于人类而言，BMI 被用来估测与肥胖相关的健康风险。但在犬猫中，目前还没有建立肥胖程度与疾病风险之间明确的联系，但在拉布拉多猎犬中发现，当脂肪量达到 25% 以上时，肥胖与胰岛素抵抗增强、骨关节炎和寿命缩短有关[58,93,94]，而猫的肥胖易于导致胰岛素抵抗和糖尿病[95,96]（请参阅第 3 章）。

4.7 总结

为了预防和治疗宠物肥胖症，简便而又可靠的体成分评估系统显得至关重要。如果无法使用 D_2O、DEXA、CT 或 MRI 等先进方法，则可以使用 5 分制或 9 分制体况评分系统，其也已被反复证明是识别犬猫肥胖的可接受的方法。然而，BCS 和平均体脂率之间的重叠评分现象与两者之间的差异，限制了该系统在宠物肥胖症研究中的适用性。此外，新开发的 BFI 可能是估测严重肥胖宠物最佳体重的首选方法。

参考文献

1. Courcier EA, O'Higgins R, Mellor DJ, Yam PS. Prevalence and risk factors for feline obesity in a first opinion practice in Glasgow, Scotland. *Journal of Feline Medicine and Surgery* 2010;12(10):746–753.
2. Courcier EA, Mellor DJ, Thomson RM, Yam PS. A cross sectional study of the prevalence and risk factors for owner misperception of canine body shape in first opinion practice in Glasgow. *Preventive Veterinary Medicine* 2011;102(1):66–74.
3. Eastland-Jones RC, German AJ, Holden SL, Biourge V, Pickavance LC. Owner misperception of canine body condition persists despite use of a body condition score chart. *Journal of Nutritional Science* 2014;3:e45.
4. Colliard L, Ancel J, Benet JJ, Paragon BM, Blanchard G. Risk factors for obesity in dogs in France. *Journal of Nutrition* 2006;136(7 Suppl): 1951S–1954S.
5. Colliard L, Paragon BM, Lemuet B, Benet JJ, Blanchard G. Prevalence and risk factors of obesity in an urban population of healthy cats. *Journal of Feline Medicine and Surgery* 2009;11(2):135–140.
6. Remillard RL, Darden DE, Michel KE, Marks SL, Buffington CA, Bunnell PR. An investigation of the relationship between caloric intake and outcome in hospitalized dogs. *Veterinary Therapeutics: Research in Applied Veterinary Medicine* 2001;2(4): 301–310.
7. Burkholder WJ. Use of body condition scores in clinical assessment of the provision of optimal nutrition. *Journal of the American Veterinary Medical Association* 2000;217(5):650–654.
8. Lund EM, Armstrong P, Kirk CA, Klausner J. Prevalence and risk factors for obesity in adult cats from private US veterinary practices. *International Journal of Applied Research in Veterinary Medicine* 2005;3(2):88–96.
9. Bjornvad CR, Kristensen AT, Jessen LR, Annerud K. Nutritional management of veterinary hospitalized patients in Denmark. *Research Abstract Program of the 25th Annual ACVIM Forum*. Seattle, WA. 2007.
10. Freeman L, Becvarova I, Cave N et al. WSAVA global nutrition assessment guidelines. *Journal of Small Animal Practice* 2011;00(June):1–12.
11. Brooks D, Churchill J, Fein K et al. 2014 AAHA weight management guidelines for dogs and cats. *Journal of the American Animal Hospital Association* 2014;50(1):1–11.
12. Speakman JR, Booles D, Butterwick R. Validation of dual energy x-ray absorptiometry (DXA) by comparison with chemical analysis of dogs and cats. *International Journal of Obesity* 2001;25(3):439–447.

13. Burkholder WJ, Thatcher CD. Validation of predictive equations for use of deuterium oxide dilution to determine body composition of dogs. *American Journal of Veterinary Research* 1998;59(8):927–937.
14. Munday HS. Assessment of body composition in cats and dogs. *International Journal of Obesity* 1994;18:S14–S21.
15. Jensen MD. Research techniques for body composition assessment. *Journal of the American Dietetic Association* 1992;92(4):454–460.
16. Jebb SA. Measurement of soft tissue composition by dual energy x-ray absorptiometry. *The British Journal of Nutrition* 1997;77(2):151–163.
17. Plank LD. Dual-energy x-ray absorptiometry and body composition. *Current Opinion in Clinical Nutrition and Metabolic Care* 2005;8(3):305–309.
18. Grier SJ, Turner AS, Alvis MR. The use of dual-energy x-ray absorptiometry in animals. *Investigative Radiology* 1996;31(1):50–62.
19. Wang ZM, Pierson RN, Jr., Heymsfield SB. The five-level model: A new approach to organizing body-composition research. *American Journal of Clinical Nutrition* 1992;56(1):19–28.
20. Pietrobelli A, Formica C, Wang Z, Heymsfield SB. Dual-energy x-ray absorptiometry body composition model: Review of physical concepts. *American Journal of Physiology* 1996;271(6 Pt 1):E941–951.
21. Heymsfield SB, Wang Z, Baumgartner RN, Ross R. Human body composition: Advances in models and methods. *Annual Review of Nutrition* 1997;17:527–558.
22. Genton L, Hans D, Kyle UG, Pichard C. Dual-energy x-ray absorptiometry and body composition: Differences between devices and comparison with reference methods. *Nutrition* 2002;18(1):66–70.
23. Tothill P, Avenell A, Reid DM. Precision and accuracy of measurements of whole-body bone mineral: Comparisons between Hologic, Lunar and Norland dual-energy x-ray absorptiometers. *The British Journal of Radiology* 1994;67(804):1210–1217.
24. Tothill P, Avenell A, Love J, Reid DM. Comparisons between Hologic, Lunar and Norland dual-energy x-ray absorptiometers and other techniques used for whole-body soft tissue measurements. *European Journal of Clinical Nutrition* 1994;48(11):781–794.
25. Sakai Y, Ito H, Meno T, Numata M, Jingu S. Comparison of body composition measurements obtained by two fan-beam DXA instruments. *Journal of Clinical Densitometry* 2006;9(2):191–197.
26. Lauten SD, Cox NR, Brawner WR, Baker HJ. Use of dual energy x-ray absorptiometry for noninvasive body composition measurements in clinically normal dogs. *American Journal of Veterinary Research* 2001;62(8):1295–1301.
27. Svendsen OL, Haarbo J, Hassager C, Christiansen C. Accuracy of measurements of body composition by dual-energy x-ray absorptiometry *in vivo*. *The American Journal of Clinical Nutrition* 1993;57(5):605–608.
28. Brunton JA, Bayley HS, Atkinson SA. Validation and application of dual-energy x-ray absorptiometry to measure bone mass and body composition in small infants. *The*

American Journal of Clinical Nutrition 1993;58(6):839–845.

29. Pintauro SJ, Nagy TR, Duthie CM, Goran MI. Cross-calibration of fat and lean measurements by dual-energy x-ray absorptiometry to pig carcass analysis in the pediatric body weight range. *American Journal of Clinical Nutrition* 1996;63(3):293–298.

30. Mitchell AD, Conway JM, Potts WJE. Body composition analysis of pigs by dual-energy x-ray absorptiometry. *Journal of Animal Science* 1996;74(11):2663–2671.

31. Picaud JC, Rigo J, Nyamugabo K, Milet J, Senterre J. Evaluation of dual-energy x-ray absorptiometry for body-composition assessment in piglets and term human neonates. *American Journal of Clinical Nutrition* 1996;63(2):157–163.

32. Ellis KJ, Shypailo RJ, Pratt JA, Pond WG. Accuracy of dual-energy x-ray absorptiometry for body-composition measurements in children. *The American Journal of Clinical Nutrition* 1994;60(5):660–665.

33. Mitchell AD, Scholz AM, Conway JM. Body composition analysis of small pigs by dual-energy x-ray absorptiometry. *Journal of Animal Science* 1998;76(9):2392–2398.

34. Black A, Tilmont EM, Baer DJ, Rumpler DK, Roth GS, Lane MA. Accuracy and precision of dual-energy x-ray absorptiometry for body composition measurements in rhesus monkeys. *Journal of Medical Primatology* 2001;30(2):94–99.

35. Son HR, d'Avignon A, Laflamme DP. Comparison of dual-energy x-ray absorptiometry and measurement of total body water content by deuterium oxide dilution for estimating body composition in dogs. *American Journal of Veterinary Research* 1998;59(5):529–532.

36. Koo WW, Hammami M, Hockman EM. Use of fan beam dual energy x-ray absorptiometry to measure body composition of piglets. *The Journal of Nutrition* 2002;132(6):1380–1383.

37. Brunton JA, Weiler HA, Atkinson SA. Improvement in the accuracy of dual energy x-ray absorptiometry for whole body and regional analysis of body composition: Validation using piglets and methodologic considerations in infants. *Pediatric Research* 1997;41(4):590–596.

38. Munday HS, Booles D, Anderson P, Poore DW, Earle KE. The repeatability of body composition measurements in dogs and cats using dual energy x-ray absorptiometry. *The Journal of Nutrition* 1994;124(12 Suppl):2619s–2621s.

39. Toll PW, Gross KL, Berryhill SA, Jewell DE. Usefulness of dual energy x-ray absorptiometry for body composition measurement in adult dogs. *The Journal of Nutrition* 1994;124(12 Suppl):2601S–2603S.

40. Lauten SD, Cox NR, Baker GH, Painter DJ, Morrison NE, Baker HJ. Body composition of growing and adult cats as measured by use of dual energy x-ray absorptiometry. *Comparative Medicine* 2000;50(2):175–183.

41. German AJ, Holden S, Bissot T, Morris PJ, Biourge V. Changes in body composition during weight loss in obese client-owned cats: Loss of lean tissue mass correlates with overall percentage of weight lost. *Journal of Feline Medicine and Surgery* 2008;10(5):452–459.

42. Munday HS, Earle KE, Anderson P. Changes in the body composition of the domestic shorthaired cat during growth and development. *The Journal of Nutrition* 1994;124(12 Suppl):2622S–2623S.
43. Mosing M, German AJ, Holden SL et al. Oxygenation and ventilation characteristics in obese sedated dogs before and after weight loss: A clinical trial. *Veterinary Journal* 2013;198(2):367–371.
44. Vitger AD, Stallknecht BM, Nielsen DH, Bjornvad CR. Integration of a physical training program in a weight loss plan for overweight pet dogs. *Journal of the American Veterinary Medical Association* 2016;248(2):174–182.
45. German AJ, Holden SL, Morris PJ, Biourge V. Comparison of a bioimpedance monitor with dual-energy x-ray absorptiometry for noninvasive estimation of percentage body fat in dogs. *American Journal of Veterinary Research* 2010;71(4):393–398.
46. Mawby DI, Bartges JW, d'Avignon A, Laflamme DP, Moyers TD, Cottrell T. Comparison of various methods for estimating body fat in dogs. *Journal of the American Animal Hospital Association* 2004;40(2):109–114.
47. Laflamme D. Development and validation of a body condition score system for cats: A clinical tool. *Feline Practice* 1997;25(5/6):13–18.
48. German AJ, Holden SL, Moxham GL, Holmes KL, Hackett RM, Rawlings JM. A simple, reliable tool for owners to assess the body condition of their dog or cat. *The Journal of Nutrition* 2006;136(7):2031S–2033S.
49. Bjornvad CR, Nielsen DH, Armstrong PJ et al. Evaluation of a nine-point body condition scoring system in physically inactive pet cats. *American Journal of Veterinary Research* 2011;72(4):433–437.
50. Michel KE, Anderson W, Cupp C, Laflamme DP. Correlation of a feline muscle mass score with body composition determined by dual-energy x-ray absorptiometry. *British Journal of Nutrition* 2011;106(Suppl 1): S57–S59.
51. Witzel AL, Kirk CA, Henry GA, Toll PW, Brejda JJ, Paetau-Robinson I. Use of a novel morphometric method and body fat index system for estimation of body composition in overweight and obese dogs. *Journal of the American Veterinary Medical Association* 2014;244(11):1279–1284.
52. Witzel AL, Kirk CA, Henry GA, Toll PW, Brejda JJ, Paetau-Robinson I. Use of a morphometric method and body fat index system for estimation of body composition in overweight and obese cats. *Journal of the American Veterinary Medical Association* 2014;244(11):1285–1290.
53. Bushberg JT. *The Essential Physics of Medical Imaging*. 3rd ed. Philadelphia: Lippincott Williams & Wilkins; 2011.
54. Ishioka K, Okumura M, Sagawa M, Nakadomo F, Kimura K, Saito M. Computed tomographic assessment of body fat in beagles. *Veterinary Radiology & Ultrasound* 2005;46(1):49–53.
55. Lee H, Kim M, Choi M et al. Assessment of feline abdominal adipose tissue using computed tomography. *Journal of Feline Medicine and Surgery* 2010;12(12):936–941.

56. Buelund LE, Nielsen DH, McEvoy FJ, Svalastoga EL, Bjornvad CR. Measurement of body composition in cats using computed tomography and dual energy x-ray absorptiometry. *Veterinary Radiology & Ultrasound* 2011;52(2):179–184.
57. Kobayashi T, Koie H, Kusumi A, Kitagawa M, Kanayama K, Otsuji K. Comparative investigation of body composition in male dogs using CT and body fat analysis software. *The Journal of Veterinary Medical Science* 2014;76(3):439–446.
58. Muller L, Kollar E, Balogh L et al. Body fat distribution and metabolic consequences: Examination opportunities in dogs. *Acta Veterinaria Hungarica* 2014;62(2):169–179.
59. Hoenig M, Thomaseth K, Waldron M, Ferguson DC. Insulin sensitivity, fat distribution, and adipocytokine response to different diets in lean and obese cats before and after weight loss. *American Journal of Physiology: Regulatory, Integrative and Comparative Physiology* 2007;292(1):R227–234.
60. Gifford A, Kullberg J, Berglund J et al. Canine body composition quantification using 3 tesla fat-water MRI. *Journal of Magnetic Resonance Imaging* 2014;39(2):485–491.
61. Zanghi BM, Cupp CJ, Pan YL et al. Noninvasive measurements of body composition and body water via quantitative magnetic resonance, deuterium water, and dual-energy x-ray absorptiometry in awake and sedated dogs. *American Journal of Veterinary Research* 2013;74(5):733–743.
62. Zanghi BM, Cupp CJ, Pan YL et al. Noninvasive measurements of body composition and body water via quantitative magnetic resonance, deuterium water, and dual-energy x-ray absorptiometry in cats. *American Journal of Veterinary Research* 2013;74(5):721–732.
63. Ballevre O, Anantharaman-Barr G, Gicquello P, Piguet-Welsh C, Thielin AL, Fern E. Use of the doubly-labeled water method to assess energy expenditure in free living cats and dogs. *Journal of Nutrition* 1994;124(12 Suppl):2594S–2600S.
64. Speakman JR, Perez-Camargo G, McCappin T, Frankel T, Thomson P, Legrand-Defretin V. Validation of the doubly-labelled water technique in the domestic dog (*Canis familiaris*). *British Journal of Nutrition* 2001;85(1):75–87.
65. Kyle UG, Bosaeus I, De Lorenzo AD et al. Bioelectrical impedance analysis—Part I: Review of principles and methods. *Clinical Nutrition* 2004;23(5):1226–1243.
66. Lukaski HC. Evolution of bioimpedance: A circuitous journey from estimation of physiological function to assessment of body composition and a return to clinical research. *European Journal of Clinical Nutrition* 2013;67(Suppl 1):S2–S9.
67. Kyle UG, Bosaeus I, De Lorenzo AD et al. Bioelectrical impedance analysis-part II: utilization in clinical practice. *Clinical Nutrition* 2004;23(6):1430–1453.
68. Stanton CA, Hamar DW, Johnson DE, Fettman MJ. Bioelectrical impedance and zoometry for body composition analysis in domestic cats. *American Journal of Veterinary Research* 1992;53(2):251–257.
69. Elliott DA, Backus RC, Van Loan MD, Rogers QR. Evaluation of multifrequency bioelectrical impedance analysis for the assessment of extracellular and total body water in healthy cats. *Journal of Nutrition* 2002;132(6 Suppl 2):1757S–1759S.

70. Elliott DA, Backus RC, Van Loan MD, Rogers QR. Extracellular water and total body water estimated by multifrequency bioelectrical impedance analysis in healthy cats: A cross-validation study. *Journal of Nutrition* 2002;132(6 Suppl 2):1760S–1762S.
71. Jeusette I, Greco D, Aquino F et al. Effect of breed on body composition and comparison between various methods to estimate body composition in dogs. *Research in Veterinary Science* 2010;88(2):227–232.
72. Kienzle E, Moik K. A pilot study of the body weight of pure-bred client-owned adult cats. *The British Journal of Nutrition* 2011;106(Suppl 1): S113–S115.
73. Shoveller AK, DiGennaro J, Lanman C, Spangler D. Trained vs untrained evaluator assessment of body condition score as a predictor of percent body fat in adult cats. *Journal of Feline Medicine and Surgery* 2014;16(12):957–965.
74. Burkholder W, Toll P. Obesity. In: Hand MS, Thatcher CD, Remillard RL, Roudebush P, editors. *Small Animal Clinical Nutrition*. 4th ed. Topeka, KS: Mark Morris Institute; 2000, pp. 401–430.
75. Butterwick R. How fat is that cat? *Journal of Feline Medicine and Surgery* 2000;2(2):91–94.
76. Hawthorne A, Butterwick R. Predicting the body composition of cats: Development of a zoometric measurement for estimation of percentage body fat in cats [abstract]. *Journal of Veterinary Internal Medicine* 2000;14(3):365.
77. Falkenberg M, Hoelmkjaer KM, Cronin A, Nielsen DH, Bjornvad CR. Evaluering af overensstemmelsen af body condition score og feline body mass index sammenlignet med dual energy x-ray absorptiometry hos katte. *Dansk Veterinaertidsskrift* 2016;4:32–37.
78. Laflamme D. Development and validation of a body condition score system for dogs. *Canine Practice* 1997;22(4):10–15.
79. Edney AT, Smith PM. Study of obesity in dogs visiting veterinary practices in the United Kingdom. *Veterinary Record* 1986;118(14):391–396.
80. Diez M. Body condition scoring in cats and dogs. *Waltham Focus* 2006;16(1):39–40.
81. Scarlett JM, Donoghue S, Saidla J, Wills J. Overweight cats: Prevalence and risk factors. *International Journal of Obesity and Related Metabolic Disorders* 1994;18(Suppl 1):S22–S28.
82. Corbee RJ. Obesity in show cats. *Journal of Animal Physiology and Animal Nutrition* 2014;98(6):1075–1080.
83. Corbee RJ. Obesity in show dogs. *Journal of Animal Physiology and Animal Nutrition* 2013;97(5):904–910.
84. Scott KC, Levy JK, Gorman SP, Newell SM. Body condition of feral cats and the effect of neutering. *Journal of Applied Animal Welfare Science* 2002;5(3):203–213.
85. Weeth LP, Fascetti AJ, Kass PH, Suter SE, Santos AM, Delaney SJ. Prevalence of obese dogs in a population of dogs with cancer. *American Journal of Veterinary Research* 2007;68(4):389–398.
86. Yaissle JE, Holloway C, Buffington CA. Evaluation of owner education as a component of obesity treatment programs for dogs. *Journal of the American Veterinary Medical Association* 2004;224(12):1932–1935.

87. Kronfeld DS, Donoghue S, Glickman LT. Body condition of cats. *Journal of Nutrition* 1994;124(12 Suppl):2683S–2684S.
88. Baez JL, Michel KE, Sorenmo K, Shofer FS. A prospective investigation of the prevalence and prognostic significance of weight loss and changes in body condition in feline cancer patients. *Journal of Feline Medicine and Surgery* 2007;9(5):411–417.
89. Laflamme D. Estimation of body fat by body condition score. *Journal of Veterinary Internal Medicine* 1994;8(2):154.
90. Cline MG, Witzel AL, Moyers TD, Bartges JW, Kirk CA. Body composition of outdoor-intact cats compared to indoor-neutered cats using dual energy x-ray absorptiometry [abstract]. *Journal of Animal Physiology and Animal Nutrition* 2013;97(6):1191.
91. Royal Canin. Body condition charts [December 11, 2015]. Available from: http://www.royalcaninhealthyweight.co.uk/pet-obesity.
92. German AJ, Holden SL, Bissot T, Morris PJ, Biourge V. Use of starting condition score to estimate changes in body weight and composition during weight loss in obese dogs. *Research in Veterinary Science* 2009;87(2):249–254.
93. Lawler DF, Larson BT, Ballam JM et al. Diet restriction and ageing in the dog: Major observations over two decades. *British Journal of Nutrition* 2008;99(4):793–805.
94. Smith GK, Paster ER, Powers MY et al. Lifelong diet restriction and radiographic evidence of osteoarthritis of the hip joint in dogs. *Journal of the American Veterinary Medical Association* 2006;229(5):690–693.
95. Sallander M, Eliasson J, Hedhammar A. Prevalence and risk factors for the development of diabetes mellitus in Swedish cats. *Acta Veterinaria Scandinavica* 2012;54:61.
96. Bjornvad CR, Rand JS, Tan HY et al. Obesity and sex influence insulin resistance and total and multimer adiponectin levels in adult neutered domestic shorthair client-owned cats. *Domestic Animal Endocrinology* 2014;47:55–64.

5 肥胖症宠物的营养管理策略

5.1 肥胖宠物评估

除了评估肥胖症宠物的体况,还有必要花时间对宠物日常饮食、营养和运动提出建议。首先要对肥胖程度进行准确诊断,然后再评估目标体重(target body weight,TBW)或理想体重(ideal body weight,IBW)。在制订饮食和减肥计划之前,应比较当前的能量摄入量和达到减肥效果的能量需要量。包括达到目标体重之后,持续地重新评估能量需要量,才能长期成功地维持犬猫的理想体重。

5.1.1 宠物肥胖症的诊断

当发现宠物肥胖时,兽医团队的任务是排除或妥善处理所有合并症(见第3章),通过宠物体成分准确诊断病情(见第4章)。当体重和体况评分(图5.1)显示宠物超重或肥胖时,宠物主人和兽医团队须能识别判定出来,并讨论针对这种疾病可行的治疗方案。针对初诊或转诊接受兽医治疗的犬猫开展的调查显示,超重或肥胖犬和猫中分别只有11.4%和3.6%的比例被认定为具有超重或肥胖的健康问题[1,2]。而且,许多主人低估了宠物的体况评分,即使那些认识到体况评分有增高趋势的宠物主人,也往往不会认为他们的宠物存在超重问题[3-5](见第6章)。

如果按照9分制体况评分(BCS)的设置,体况评分每增加1相当于体脂率(body fat,BF)增加了约5%[6,7](表5.1)。猫和犬的BCS达到6~7/9(BF为25%~34%)即视为达到超重,如果BCS ≥ 8/9则视为达到肥胖(BF ≥ 35%)。体况评分(BCS)系统只涵盖到BF最高为45%的动物,但有研究报道,猫的BF最高可达62%,犬的BF最高可达65%[8,9]。按9分制的体况评分表,每增加1分BCS意味着体重增加约10%(表5.1)[6,7,10]。由于超重动物的体重可能超过40%,或体脂率(BF)可能超过45%,因此,除了体况评分方法外,宠物主人和兽医团队还需要使用其他工具,以帮助他们更准确地诊断宠物肥胖症。体脂指数(body fat index,BFI)(图5.2)的设计目的是通过触诊方式,使用具体测定的数据描述来确定评估的体脂率,其所测得的动物BF最高可达65%[8,9](见第4章)。在检查室使用这些工具并邀请宠物主

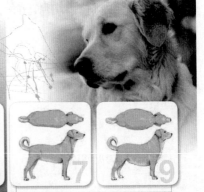

体况评分

低于理想状态
① 从一定距离观察，肋骨、腰椎、盆骨和所有骨骼突起明显。无可视脂肪存在，肌肉量明显缺少
② 容易看到肋骨、腰椎和盆骨。无可触及脂肪。其他骨骼有一些突起。肌肉轻微减少
③ 肋骨容易触及且可视，无可触及脂肪。可见腰椎顶部。盆骨突出。腰部和腹部有明显皱褶

理想状态
④ 肋骨容易触及，脂肪覆盖率极低。从上方观察，腰部很容易看出。腹部有明显皱褶
⑤ 肋骨可触及且无多余脂肪覆盖。从上方观察时，在肋骨后面可观察到腰部。侧面观察腹部收拢

高于理想状态
⑥ 肋骨可触及，有轻微的多余脂肪覆盖，从上面观察可以看出腰部，但并不明显。腹部皱褶可见
⑦ 肋骨难以触及，有厚厚的脂肪覆盖。腰部和尾巴基部有明显的脂肪沉积。腰部不可见或几乎看不到。腹部皱褶可能看得见
⑧ 肋骨覆盖过多脂肪无法触及，或者只有在很大的压力下才能摸到。大量脂肪沉积在腰部区域和尾巴基部。腰部不可见。没有腹部皱褶。腹部可能出现明显膨大
⑨ 大量脂肪沉积在胸部、脊柱和尾巴基部。腰部和腹部没有皱褶。颈部和四肢脂肪沉积。腹部明显膨大

资料来源：

(a)

体况评分

   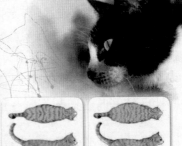

低于理想状态
① 短毛猫可见肋骨。无可触及脂肪。腹部皱褶极多。腰椎和髂骨翼容易触及
② 短毛猫很容易看到肋骨。腰椎明显。腹部皱褶明显。无可触及脂肪
③ 肋骨容易触及，有少量脂肪覆盖。腰椎明显。肋骨后腰明显。腹部少量脂肪

理想状态
④ 肋骨可触及，有少量脂肪覆盖。腰部以下有明显的赘肉。腹部略有皱褶。无腹部脂肪垫
⑤ 体型匀称。可观察到肋骨后腰部，肋骨可触及，有轻微脂肪覆盖。腹部少量脂肪垫

高于理想状态
⑥ 肋骨可触及，并有少量脂肪覆盖。腰部和腹部脂肪垫可辨别但不明显。腹部无皱褶
⑦ 肋骨不容易触及，有中度脂肪覆盖。腰部不明显。腹部明显变圆。中度腹部脂肪垫
⑧ 肋骨不可触及，且有多余脂肪覆盖。腰部不可见。腹部明显变圆。腹部脂肪垫突出。腰部出现脂肪沉积
⑨ 肋骨不可触及，且覆盖大量脂肪。腰部、面部和四肢有大量脂肪沉积。腹部膨大，无法看到腰部。腹部大量脂肪堆积

资料来源：

(b)

图 5.1 犬和猫的体况评分系统（9 分制）（世界小动物兽医协会全球营养委员会）

人参与对宠物的触诊,可有效帮助主人了解其宠物的超重或肥胖状况,并进一步确诊宠物肥胖的状况。

表 5.1　体况评分与体脂率及超重的关系

体况评分(9 分制)	体脂率 /%	超重 /%
4	15~19	
5	20~24	
6	25~29	10
7	30~34	20
8	35~39	30
9	40+	40

来源:Brooks Detal., *Journal of the American Animal Hospital Association* 2014;50:1-11.

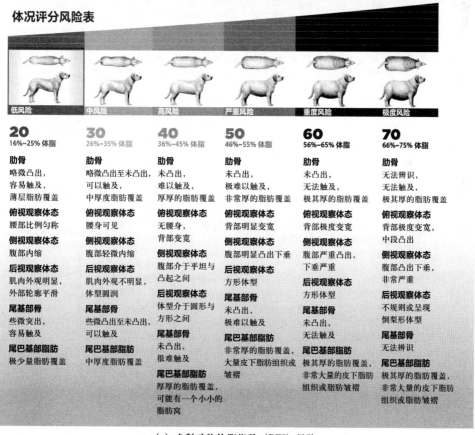

(a) 犬科动物体脂指数(BFI)风险

图 5.2　犬科动物和猫科动物的体脂指数(参考 Hill's Pet Nutrition,Inc)

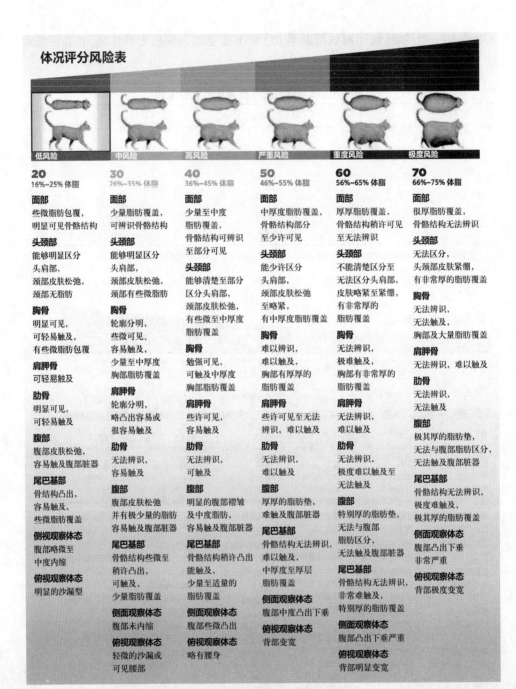

(b) 猫科动物体脂指数（BFI）风险

续图 5.2

5.1.2 估算目标体重或理想体重

体重管理的一个重要部分是为肥胖宠物确定目标体重 TBW 或理想体重 IBW。目标体重 TBW 代表的是兽医团队认为宠物需要达到的体重,是减肥计划的一部分。如果达到理想体重 IBW 所需减掉的体重量是不合适的,则这二者是有区别的。这种情况通常是由肥胖合并症如心血管疾病或肾脏疾病引起。当宠物医疗记录同时包含体重和体况评分 BCS 数据时,则可以根据历史数据确定理想体重。还可以使用体脂率、体重超重百分比或体况评分 BCS 的方程式来估算理想体重(框 5.1)[11,12]。临床上评估体脂率 BF 的方法包括体况评分 BCS、体脂指数 BFI 或形态测量[6-9,13]。使用形态测量预测方程式可用于预测体脂肪量、瘦体重和体脂率(框 5.2)[8,9]。作者通常使用形态测量法测量比较温顺且肥胖的犬猫。

框 5.1 估算理想体重的方程[11,12]

方程 5.1:
$$\text{理想体重(kg 或 lb)} = \frac{\text{当前体重(kg 或 lb)} \times (100\% - \text{当前体脂肪}\%)}{100\% - \text{理想体脂肪}\%}$$

方程 5.2:
$$\text{理想体重(kg 或 lb)} = \frac{\text{当前体重(kg 或 lb)} \times (100\% - \text{超标体重}\%)}{100}$$

方程 5.3:
$$\text{理想体重(kg 或 lb)} = \frac{\text{当前体重(kg 或 lb)} \times 100}{100 + (\text{当前 BCS} - 5) \times 10}$$

示例:10 岁雌性绝育拉布拉多犬(FS),体重 48 kg(105.6 lb),BCS 9/9,估计体脂肪为 45%

方程 5.1:48 kg(105.6 lb) × (100% − 45%)/(100% − 20%) = 33 kg(72.6 lb)

方程 5.2:48 kg(105.6 lb) × (100% − 40%)/100 = 28.8 kg(63.4 lb)

方程 5.3:48 kg(105.6 lb) × 100/[100 + (9 − 5) × 10] = 34.3 kg(75.4 lb)

框 5.2 体脂率(BF)的形态计量预测方程[8,9]

犬

BF = 0.71 × 胸围 − 0.1 × (骨盆周长/6)^2 − 5.78 × 后肢长$^{0.8}$ + 26.56 × (骨盆围/头围) + 2.06

注意:度量单位为厘米。

值得注意的是,体况评分 BCS 和体脂指数 BFI 反映的是体脂肪而不是瘦体重。随着动物年龄的增长,它们可能会出现肌肉衰减综合征(肌少症),即随着年龄增长肌肉量逐渐减少,这与潜在的炎症或病理状况无关[14]。在老年犬中,相比于颞肌(temporal muscle)和股四头肌(quadriceps muscle),轴上肌系的肌减少症表现更为明显[15]。肌肉状况评分(MCS)是世界小动物兽医协会(World Small Animal Veterinary Association,WSAVA)推荐的完整营养评估中的一部分,包括颞骨、肩胛骨、肋骨、腰椎和盆骨的触诊和可视化的方法(图 5.3)[16]。尽管这一系统还未能证实适用于瘦体重,尤其是与肥胖症相关的瘦体重,但如果是由同一个操作人员在一段时间内使用该系统效果更佳,则可以被用来主观地评估肌肉损失并具有良好的可重复性[17]。由于瘦体重的百分比影响体脂率,因此对于由年龄引起,肌肉在一定程度上萎缩的宠物,其理想的体脂率为 25% 也是可以接受的。

5.2 体重管理的能量需要

5.2.1 确定当前热量的摄入量

为了成功地管理超重或肥胖症宠物,必须评估其当前的能量摄入量。临床兽医应特别询问宠物主有关宠物摄入的每一种食物,以确保考虑到所有的能量摄入来源(框 5.3)。宠物主人可以填写一份全面的饮食、活动和家庭生活史表格(DAHHF,见第 8 章,附录 8.1),兽医可以使用该表格计算当前的能量摄入量。由于需要花费一定的时间准确填写表格,宠物主人可以在预约讨论减肥问题之前或确定宠物超重或肥胖后填写这个表格。

框 5.3　伴侣动物能量摄入来源
主要饮食:商业或家庭准备
商业或家庭自制的零食
餐桌残羹
药物管理食品
补充剂
野生动物
粪便
垃圾
其他宠物的食物

肌肉状况评分

肌肉状况评分通过对脊柱、肩胛骨、头骨和髂翼的可视化和触诊来实现。肌肉损失通常首先出现在脊柱两侧的肩胛肌；其他部位的肌肉损失可能更大。肌肉状况评分分为正常、轻度、中度、重度。注意，如果动物超重（身体状况评分 >5），它们也可能会有明显的肌肉损失。相反，动物的身体状况得分较低（<4），但肌肉损失可能会很小。因此，在每次诊疗时评估每只动物的体况评分和肌肉状况评分都很重要。当肌肉损失程度较小和动物超重时，触诊尤其重要。每个分数的示例如下所示。

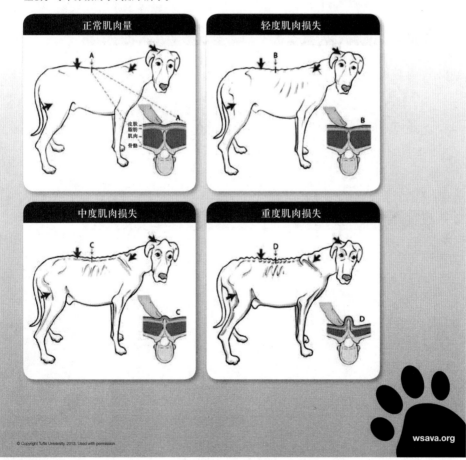

(a)

图 5.3 犬(a)和猫(b)肌肉状况评分系统
（由世界小动物兽医协会提供的全球营养委员会工具包）

肌肉状况评分

  肌肉状况评分通过对脊柱、肩胛骨、头骨和髂翼的可视化和触诊来实现。肌肉损失通常首先出现在脊柱两侧的肩胛肌；其他部位的肌肉损失可能更大。肌肉状况评分分为正常、轻度、中度、重度。注意，如果动物超重（身体状况评分 >5），它们也可能会有明显的肌肉损失。相反，动物的身体状况得分较低（<4），但肌肉损失可能会很小。因此，在每次诊疗时评估每只动物的体况评分和肌肉状况评分都很重要。当肌肉损失程度较小和动物超重时，触诊尤其重要。每个分数的示例如下所示。

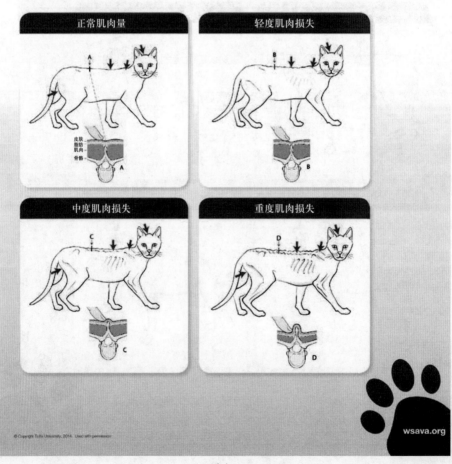

(b)

续图 5.3

市售宠物食品的能量含量可以通过参考标签、联系生产商或利用在线资源获得。根据美国饲料管理协会(Association of American Feed Control Officials, AAFCO)官方出版物的 PF9 标准法案和法规,包括零食、饮食和营养补充剂在内的犬或猫食品的标签必须标明能量含量[18]。一些咀嚼物和骨头,包括蹄子、耳朵、牛鞭和韧带等不受此标签约束(不需标明能量含量的标签),除非制造商在标签上声明其营养价值。欧洲宠物食品工业联合会(FEDIAF)目前并没有对欧洲宠物食品能量含量标签作要求。大多数人类食品的能量含量可以在美国农业部的营养数据库中找到[19]。一旦确定了总能量摄入量,就可以与估计的肥胖宠物能量需要量进行比较。

为确保当前尽可能准确地估计能量摄入量,并提供最佳喂养建议,兽医应了解宠物主人喂食时使用的宠物食具和称量器具的具体信息。虽然有时获得准确的饮食史的确很困难,但也要尽力而为。幸运的是,大多数犬主人在其宠物减肥之前所采取的喂养方式,都未影响宠物体重管理计划的成功实施[20]。如果兽医没能获得宠物完整的饮食史,仍然可以根据估计的能量需要量实施减肥计划。

5.2.2　确定减肥的能量摄入量

有几种方法可以估算减肥的初始能量摄入量(框 5.4)[11,21-23]。对于理想体重(IBW)在 2 kg 以下或 25 kg 以上的肥胖宠物,不建议使用线性方程式计算其静息能量需要(resting energy requirements, RER),因为这会高估它们的能量需要(图 5.4)。可根据理想体重(IBW)或目标体重(TBW)的具体因素,调整计算得出静息能量需要(RER)或维持能量需要(maintenance energy requirement, MER)。据报道,犬实现减肥所需的能量摄入量为 $(63 \pm 10.2) \text{ kcal}/(\text{TBW}_{kg}^{0.75} \cdot d)$[22]。猫减肥所需的能量摄入量为 $(52 \pm 4.9) \text{ kcal}/(\text{TBW}_{kg}^{0.711} \cdot d)$[23]。

框 5.4　用于减肥的静息能量需要(RER)和维持能量需要(MER)的计算[11,21-23]

计算方式 1(犬和猫):

静息能量需要(RER)

　指数方程:$\text{RER} = \text{体重}_{kg}^{0.75} \times 70$

　线性方程:$\text{RER} = 30 \times \text{体重}_{kg} + 70$

　　不适用于体重 < 2 kg 或 > 25 kg 的动物

维持能量需要(MER)

　犬:$\text{MER} = \text{RER} \times 1.0$

　猫:$\text{MER} = \text{RER} \times 0.8$

计算方式 2(仅适用于犬):
根据活动水平进行调整的选项

不活跃(<7 250 步/d)

MER = 体重$_{kg}^{0.75}$ × 95 × 0.45

活跃(>7 250 步/d)

MER = 体重$_{kg}^{0.75}$ × 95 × 0.55

计算方式 3(仅适用于犬):
基于性别和去势动物

去势母犬:MER = 体重$_{kg}^{0.75}$ × 60

母犬:MER = 体重$_{kg}^{0.75}$ × 70

去势公犬:MER = 体重$_{kg}^{0.75}$ × 70

公犬:MER = 体重$_{kg}^{0.75}$ × 80

计算方式 4(仅适用于猫):

MER = 体重$_{kg}^{0.711}$ × 53

示例:5 岁雌性绝育比格犬,估计理想体重 8.2 kg(18 lb),被认为是不活跃的

计算方式 1:

8.2 kg$^{0.75}$ × 70 = 339 kcal × 1.0 = 339 kcal/d

计算方式 2:

8.2 kg$^{0.75}$ × 95 = 460 kcal × 0.45 = 207 kcal/d

计算方式 3:

8.2 kg$^{0.75}$ × 60 = 291 kcal/d

示例:8 岁雌性美国本地短毛猫,估计理想体重 3.7 kg(8.14 lb)

计算方式 1:

3.7 kg$^{0.75}$ × 70 = 187 kcal × 0.8 = 149 kcal/d

计算方式 4:

3.7 kg$^{0.711}$ × 53 = 134 kcal/d

如果当前实际能量摄入量少于为减肥而计算出的能量摄入量,建议在此基础上将能量摄入量减少 10%~20%。一般情况下,作者不建议将犬的能量摄入量限制在 42.6 kcal/(TBW$_{kg}^{0.75}$·d)以下,不建议将猫的能量摄入量限制在 42.2 kcal/(TBW$_{kg}^{0.711}$·d)以下(约 60% RER)[22,23]。如果要求这种程度的热量限制,建议筛查宠物是否患有

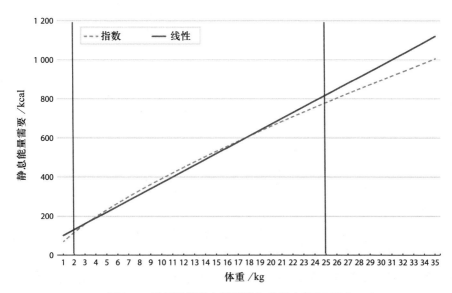

图 5.4 用于计算静息的指数与线性方程(RER)

其他混杂的合并症(请参阅第 3 章)或者咨询那些具备委员会认证的兽医营养师(Board Certified Veterinary Nutritionist™),以避免营养不足而引起潜在并发症。一项研究评估了在能量摄入限制情况下,进食 3 种非处方(over-the-counter, OTC)和 2 种体重管理处方粮,对犬营养缺乏理论上的风险评估[24]。研究发现,当能量摄入为 79 kcal/($kg^{0.75} \cdot d$)时,饮食中至少有 1 种必需营养素缺乏,而当能量摄入为 70 kcal/($kg^{0.75} \cdot d$)时,饮食中则会多种营养素缺乏。与非处方粮相比,体重管理处方粮在各限制水平下的营养缺乏较少见,最常见的营养缺乏是胆碱和硒元素。在一项后续研究中,在按照减肥计划进行减肥饮食治疗的情况下,记录到犬的血浆胆碱水平降低。然而,没有发现胆碱缺乏的迹象,而且大多数营养素没有显著变化,犬在整个减肥计划中保持着明显的健康状态[25]。

对于已经掌握精确的饮食历史并已知能量消耗量的宠物,一些职业宠物医师可能会选择将动物当前的能量摄入量减少 20%,以避免由能量摄入的突然大幅减少而引起的不适。在这种情况下,如果宠物在减重前体重还在增加,那么减重速度可能会很慢或不成功。由于这种方法可能较为保守,因此需要宠物主做好长期坚持和接受失败挫折的思想准备。在随后的治疗中,宠物体重成功降低是宠物主人坚持实施减肥计划的最好动力,在更典型的能量限制减肥计划中发现,宠物主人主观记录的犬觅食的次数会减少。然而对于猫而言,在摄入有限能量水平的食物时,则会出现觅食次数增加或减少的现象[23,26,27]。

无论用于确定理想体重/目标体重(IBW/TBW)和计算能量需要的方法如何,这些都是初始估计,都可能会随时间而发生变化。有必要进行随访以根据宠物的新陈代谢需要量身定制减肥计划,并重新评估其身体状况以确定何时达到目标体重。

5.3 实施治疗方案

5.3.1 食物选择

5.3.1.1 常量营养素

建议在明显限制能量摄入量的情况下,为肥胖动物提供体重管理处方粮。这些类型的饮食富含蛋白质,在减肥过程中可产生饱腹感并保持瘦体重,同时避免因限制能量摄入而导致蛋白质缺乏。特别是对犬来说,治疗性体重管理处方粮含有高水平的纤维,降低能量密度的同时以获得额外的饱腹感。这些饮食提高了微量营养素含量,以避免因热量限制而可能出现的潜在营养缺乏。

蛋白质是最值得关注的宏观营养素,因为蛋白质合成和糖异生需要消耗能量,这能够增加能量消耗。尽管控制了总体能量摄入量,也可以提供足够高水平蛋白质以维持充足的蛋白质摄入量[28,29]。饲喂比格犬能量含量为最低能量要求 50% 的高蛋白饮食(103 g/1 000 kcal 蛋白质)可以使其体重减轻,并恢复到与变胖前相似的体重和瘦体重水平[30]。与低蛋白质饮食的犬相比,体重超重的犬在接受热量限制同时喂食高水平的蛋白质(57~111 g/1 000 kcal)饲粮,其肌肉组织的损失变得更少[31]。美国国家研究委员会(National Research Council,NRC)推荐犬每天的蛋白质摄入量为 $3.28\ \text{g/BW}_{kg}^{0.75}$,猫为 $4.96\ \text{g/BW}_{kg}^{0.67}$ [32]。最近,有人建议猫需要 $7.8\ \text{g/BW}_{kg}^{0.75}$ 的蛋白质来维持瘦体重[33]。因此,在能量限制的情况下,为确保在减肥计划中选择的饮食能够维持足够的蛋白质摄入量,应计算蛋白质需要量(框 5.5)。

框 5.5　每日蛋白质摄入量的计算[32,33]

计算方式 1(犬):

蛋白质推荐量(RA) = $3.28\ \text{g/BW}_{kg}^{0.75}$

必需的膳食蛋白质含量 = $\dfrac{\text{蛋白质推荐量}}{\text{日能量摄入量}} \times 1\ 000$

计算方式 2(猫):

蛋白质推荐量(RA) = $4.96\ \text{g/BW}_{kg}^{0.67}$

必需的膳食蛋白质含量 = $\dfrac{\text{蛋白质推荐量}}{\text{日能量摄入量}} \times 1\ 000$

计算方式 3(猫):

维持瘦体重的蛋白质需要量 = $7.8\ \text{g/BW}_{kg}^{0.75}$

> 必需的膳食蛋白质含量 = $\dfrac{\text{蛋白质推荐量}}{\text{日能量摄入量}} \times 1\,000$
>
> 示例：5 岁雌性绝育比格犬，估计理想体重 8.2 kg(18 lb)，每天喂 339 kcal
>
> 计算方式 1：
>
> $8.2\ \text{kg}^{0.75} \times 3.28 = \dfrac{15.9\ \text{g 蛋白质}}{339\ \text{kcal}} \times 1\,000 = 46.9\ \text{g}/1\,000\ \text{kcal 蛋白质}$
>
> 示例：8 岁绝育美国短毛猫，估算理想体重 3.7 kg(8.14 lb)，每天饲喂 149 kcal
>
> 计算方式 2：
>
> $3.7\ \text{kg}^{0.67} \times 4.96 = \dfrac{11.9\ \text{g 蛋白质}}{149\ \text{kcal}} \times 1\,000 = 80\ \text{g}/1\,000\ \text{kcal 蛋白质}$
>
> 计算方式 3：
>
> $3.7\ \text{kg}^{0.75} \times 7.8 = \dfrac{20.8\ \text{g 蛋白质}}{149\ \text{kcal}} \times 1\,000 = 140\ \text{g}/1\,000\ \text{kcal 蛋白质}$

除了蛋白质含量高以外，许多减肥用处方粮还利用高纤维（high fiber, HF）含量来降低能量含量，并容易产生饱腹感。与低纤维饮食（39.1 g/1 000 kcal TDF）相比，高纤维饮食（92.5 g/1 000 kcal TDF）能够减少能量摄入。这种效应是由于高纤维饮食比低纤维饮食（3 138 kcal/kg vs. 3 615 kcal/kg ME）的能量密度低导致的，并不是饮食摄入量改变而导致的[34]。在研究低纤维和中纤维饮食时，得到同样的结果，而当每天额外进食高纤维膳食时，食物摄取量也会减少[35]。当针对犬实施能量限制的主动减肥计划时，在饮食中的全部日粮纤维（TDF）水平从 12.9 g/1 000 kcal 增加到 99.1 g/1 000 kcal 的情况下，每天的额外自由采食量没有变化。据报道，不管如何喂养，瘦的家养犬比肥胖犬摄入的粗纤维更多[37]。当猫食用不同纤维含量的饮食（22~56 g/1 000 kcal TDF）时，它们会增加食物摄入量以维持大致相同的总能量消耗，但如果额外提供美味、高能量的食物，猫食用高纤维食物则消耗的量更大[38]。其他研究表明，当猫粮中添加高岭土或不同水平的纤维素时，猫的能量摄入会减少，食物适口性下降可能是产生这种结果的主要原因[39-41]。

一项针对犬的研究，设计 3 种饮食，分别为：高蛋白质高纤维饮食（HPHF：蛋白质 103 g/1 000 kcal，粗纤维 60 g/1 000 kcal，总膳食纤维 97 g/1 000 kcal，代谢能 2 900 kcal/kg）；中蛋白质高纤维饮食（MPHF：蛋白质 86 g/1 000 kcal，粗纤维 87 g/1 000 kcal，膳食总纤维 132 g/1 000 kcal，代谢能 2 660 kcal/kg）；高蛋白质中纤维膳食（HPMF：蛋白质 104 g/1 000 kcal，粗纤维 35 g/1 000 kcal，膳食纤维 56 g/1 000 kcal，代谢能 3 275 kcal/kg）。与 HPMF 和 MPHF 两组相比，HPHF 组犬摄入的热量更

少[42]。当喂食 HPHF 饮食时，超重或肥胖的犬在主动减重计划下的体重下降率(31.8% vs. 20.0%)比 HPMF 组的犬更快(1.0%/周 vs. 0.7%/周)。与 HPMF 相比，HPHF 组犬的体脂质量损失也有更大的变化(58% vs. 37%)[43]。作为控制性减肥试验的一部分，当肥胖猫食用含有 115~139 g/1 000 kcal 蛋白质和 28~81 g/1 000 kcal 全部日粮纤维(TDF)的罐装食品或干粮时，随着时间的推移，所有猫的体重和体况评分(BCS)都会下降，但主人们最不满意的是，当只喂含有 115 g/1 000 kcal 蛋白质和 81 g/1 000 kcal 全部日粮纤维的干粮时，他们的猫饥饿感更强烈[26]。

尤其是对猫而言，一些减轻体重的喂养建议，应为宠物提供低碳水化合物(low carbohydrate, LC)饮食而不是增加膳食纤维含量。低碳水化合物饮食的目的是使机体处于糖原分解状态，当糖原储备耗尽时，机体动用脂肪进行 β-氧化产生替代能源。同时，乙酰辅酶 A 的增加导致肝酮体(丙酮、乙酰乙酸、β-羟基丁酸)随之增加，为大多数组织提供替代能源，并为葡萄糖依赖组织(如脑和红细胞)保留葡萄糖。葡萄糖的转出会导致空腹和餐后血糖及循环胰岛素降低。这些类型的饮食可能比高纤维 HF 饮食更具有代谢优势，特别是对于因肥胖和胰岛素抵抗而患糖尿病的猫[44]，无论选择何种饮食，减轻体重对于控制或缓解猫的糖尿病是最为重要的[45]。

低脂肪饮食并不是专门用于犬或猫的减肥方案。除了低碳水化合物 LC 饮食，许多犬和猫的处方粮倾向于采用较低的脂肪含量，但是，这只能作为一种降低饮食中能量密度的方法，而不是因为其对于减脂有明显的效果。

5.3.1.2 饮食形式和能量密度

饮食选择应同时考虑肥胖宠物的饮食偏好(干的还是罐装的)和宠物主人的喜好。与高纤维 HF 减重处方粮相比，干的低碳水化合物 LC 饮食往往具有高能量密度，太少量的食物往往导致肥胖宠物或其主人变得很难坚持下去。例如，体重为 5 kg 的猫每天可能需要消耗 190 kcal 热量来减肥，这还不到目前可用的治疗性干的低碳水化合物 LC 饮食 8 盎司量杯标准的 1/3(磨碎约 45 g)。罐头食品可能通过提高水分含量来增加猫的饱腹感并达到体重减轻的目的。食用罐装食品的猫与食用相同组分冻干粮的猫相比，前者能量摄入量和体重明显下降，这表明在猫的减肥过程中，较高水分的食物可能比干的食物更有益[46]。

成功的体重管理可以通过低碳水化合物 LC 和高纤维 HF 饮食来实现，但是很少有研究对比过猫饲喂 LC 饮食和饲喂 HF 饮食的效果差异。在一项研究中表明，宠物身体状况和摄入的能量(而不是营养成分)在减肥过程中影响了体重变化[47]。另一项研究表明，猫主人评价正在减肥的宠物猫的饱腹感未发现 LC 和 HF 饮食之间的差异，然而这项研究的样本量有限[48]。如果首选干粮，作者通常推荐高蛋白质、低能量密度的食物(通过添加纤维和/或减少脂肪)，因为低碳水化合物 LC 饮食中往往脂肪含量较高，从而导致能量密度较高。

5.3.2 减肥食物分配

减肥食品应通过罐/袋/托盘称重量或用标准量杯称量。最好使用重量来测量干的膨化食物,因为使用量杯会导致精度和准确性下降(分量低估 18% 至高估 80% 不等),而精度与分量呈负相关[49]。相对于小碗和小勺,犬主人在使用大碗和大勺时更有可能给宠物饲喂更多的犬粮[50]。有的主人可能使用量杯,当然也可能是使用无标记的通用勺子,无意中提供了比预期更多的食物,这将导致食物消耗报告并不准确。在厨房秤上称食物的重量(以克为单位)对猫或幼犬特别有用,因为测量杯子容积的变化可能会导致测量的食物热量有显著差异。可以在每个产品标签上或通过制造商获得以 kcal/kg 为单位的热量信息。改用较小的宠物食具也可以帮助一些宠物主人减轻错觉,他们总觉得给宠物的食物分量不足[50]。

5.3.3 制定饮食指南

对于任何宠物来说,一般建议除了摄取全面而均衡的饮食外,其他食物所摄取的能量不得超过总能量的 10%,以避免饮食中营养物质被不完整且不均衡的食物稀释。这包括零食、餐桌上的食品、用于给药的食物,以及来自补充剂的热量。建议使用低能量的水果和蔬菜以帮助宠物主人坚持喂养计划(框 5.6)。在选择商业零食时,应确定热量含量,并应控制每天食用零食的数量。每天使用为宠物准备好的治疗盒或治疗袋是一种有效的喂养策略,以避免发生犬被过度喂食的情况。每日饮食限额表可以帮助宠物主人更好地利用他们的饮食限额,避免超额供应食物(见第 8 章,附录 8.3)。

框 5.6 低热量水果和蔬菜饮食限量表[19]

项目	通用单位	1 杯 (kcal)	100 g (kcal)
生的小胡萝卜	1 个大的 = 5 kcal		35
生的绿豆	4 颗 = 2 kcal	62	30
生的青椒	10 条 = 5 kcal	18	20
生芹菜	5 柄 = 2 kcal	16	16
切片西葫芦(蒸/煮)	1 个中等大小 = 33 kcal	28	15
生的带皮苹果	1 个大的 = 116 kcal	65	52
生的蓝莓	1 颗 = 1 kcal	84	57
生的西瓜	1 球 = 4 kcal	46	30
罐装南瓜	1 汤匙 = 5 kcal	83	34

5.4 重新评估减肥计划

5.4.1 例行随访

应在减肥计划开始后2~4周进行首次的减肥回访(见第8章)。在每次回访中,计算减重率(框5.7),记录减肥计划开始以来的饮食史,调整食物摄入量,并记录是否坚持按计划方案执行,处理那些没有坚持计划的问题。另外,评估体况评分(BCS)和肌肉状况评分。建议每周减重率为1%~2%,但对猫来说减重率为0.5%~1%也是可以接受的。如果肥胖宠物减轻的重量在此可接受的范围内,则不需要更改饮食计划,除非宠物主人提出无法继续坚持这个方案了。

框5.7 计算每周减肥率

$$总减肥率 = \frac{减肥前体重 - 当前体重}{减肥前体重} \times 100$$

$$每周减肥率 = \frac{总减肥率}{自上次称重后周数}$$

示例:8岁雌性美国本地短毛猫,减肥前体重5 kg(11 lb),现在的体重为4.9 kg(10.8 lb),估算理想体重3.7 kg(8.14 lb),距离上次体重称量2周。

总减肥率 = (5 kg − 4.9 kg)/5 kg × 100 = 2%

每周减肥率 = 2%/2 = 1%

对于比率超出此可接受范围或并没有严格遵守减肥计划的肥胖宠物,应详细评估肥胖宠物当前食物摄入量并与推荐的减肥计划做比较。需要再次强调宠物需要分开喂养,要求宠物明确列出正在喂食的食物,每种食物的体积或重量,以及是否有其他在计划饮食之外的喂食。这项检查很可能就确定了之后提到的饲喂行为的改变(见第6章)。如果减肥率超出可接受的范围,并且没有发现不严格遵守方案的问题,则应相应地调整食物摄入量。热量通常减少5%~10%(包括零食),并且每2~4周进行一次体重检查,直到体重持续减轻为止。在肥胖宠物减肥进展良好的情况下,可以考虑延长2次体重检查的间隔时间,但最好至少每个月称重1次。

5.4.2 保持长期成功效果

根据肥胖的程度,成功而安全的减肥可能需要几个月,甚至可能超过1年。在12周内,只有0.6%的犬保持了每周2%的减重率,平均达到每周初始体重的(0.9 ± 0.45)%。在同一时间段内,减肥率也从每周平均1.3%降至每周0.8%[22]。在一项类似的研究中,猫每周减重率(0.8 ± 0.50)%,从每周1.2%降至0.7%[23]。

一旦达到理想体重或目标体重时(IBW或TBW),就可以调整能量摄入以维持当前的体重。减重后,犬的维持能量需要为(52~104)kcal/kg$^{0.75}$,一般只比减重时多10%左右[51]。尽管在猫上缺乏类似的数据,但建议猫和犬在维持体重期间饮食的热量增加不超过10%。维持饲喂调整后,需要额外的跟踪随访(通常要进行1~3次复查),以确保维持理想/目标体重。减肥成功后,如果由犬来自由选择食物,那么它们的体重会反弹[52]。犬也会比之前变胖时消耗更少的热量,却更快地反弹到之前的肥胖水平[53]。

如果改用标准的维持饮食,犬会消耗更多的热量并反弹更多的体重,这表明长期饲喂定制体重管理目标的饮食可以明显控制其体重反弹[54]。然而,一部分肥胖的犬和猫在达到了理想体重或目标体重后,体重又在此基础上增加了5%,犬和猫中这一比例分别是49%和46%。与犬相比,猫体重的反弹可能会超过之前减掉的50%体重。虽然这些数据可能令人沮丧,但重要的是要记住,大多数的犬和猫(51%~54%)在成功减肥后维持住了体重,甚至还有的体重继续减轻。

5.5 结论

尽管宠物的减肥计划在执行过程中可能会遇到令人沮丧的障碍,但重要的是要提醒宠物主人,只要他们愿意付出必要的时间和努力,就可以让宠物长期保持体重,这是可以做到的。建议长期坚持低能量密度的体重管理饮食,给予宠物特定的治疗和锻炼,以避免体重反弹。

参考文献

1. Lund EM, Armstrong PJ, Kirk CA, Klausner JS. Prevalence and risk factors for obesity in adult cats from private US veterinary practices. *The International Journal of Applied Research in Veterinary Medicine* 2005;3:88–96.
2. Lund EM, Armstrong PJ, Kirk CA, Klausner JS. Prevalence and risk factors for obesity in adult dogs from private US veterinary practices. *The International Journal of Applied Research in Veterinary Medicine* 2006;4:177–186.
3. Rohlf VI, Toukhsati S, Coleman GJ, Bennett PC. Dog obesity: Can dog caregivers' (owners') feeding and exercise intentions and behaviors be predicted from attitudes? *Journal of Applied Animal Welfare Science* 2010;13:213–236.
4. White GA, Hobson-West P, Cobb K, Craigon J, Hammond R, Millar KM. Canine obesity: Is there a difference between veterinarian and owner perception? *Journal of Small Animal Practice* 2011;52:622–626.

5. Rowe EC, Browne WJ, Casey RA, Gruffydd-Jones TJ, Murray JK. Early- life risk factors identified for owner-reported feline overweight and obesity at around two years of age. *Preventive Veterinary Medicine* 2017;14:39–48.
6. Laflamme D. Development and validation of a body condition score system for cats: A clinical tool. *Feline Practice* 1997;25:13–17.
7. Laflamme D. Development and validation of a body condition score system for dogs. *Canine Practice* 1997;22:10–15.
8. Witzel AL, Kirk CA, Henry GA, Toll PW, Brejda JJ, Paetau-Robinson I. Use of a novel morphometric method and body fat index system for estimation of body composition in overweight and obese dogs. *Journal of the American Veterinary Medical Association* 2014;22:1279–1284.
9. Witzel AL, Kirk CA, Henry GA, Toll PW, Brejda JJ, Paetau-Robinson I. Use of a morphometric method and body fat index system for estimation of body composition in overweight and obese cats. *Journal of the American Veterinary Medical Association* 2014;24:1285–1290.
10. Mawby DI, Bartges JW, d'Avignon A, Laflamme DP, Moyers TD, Cottrell T. Comparison of various methods for estimating body fat in dogs. *Journal of the American Animal Hospital Association* 2004;40:109–114.
11. Brooks D, Churchill J, Fein K et al. 2014 AAHA weight management guidelines for dogs and cats. *Journal of the American Animal Hospital Association* 2014;50:1–11.
12. German AJ, Holden SL, Bissot T, Morris PJ, Biourge V. Use of starting condition score to estimate changes in body weight and composition during weight loss in obese dogs. *Research in Veterinary Science* 2009;87:249–254.
13. Bjornvad CR, Nielsen DH, Armstrong PJ et al. Evaluation of a nine-point body condition scoring system in physically inactive pet cats. *American Journal of Veterinary Research* 2011;72:433–437.
14. Freeman LM. Cachexia and sarcopenia: Emerging syndromes of importance in dogs and cats. *Journal of Veterinary Internal Medicine* 2012;26:3–17.
15. Hutchinson D, Sutherland-Smith J, Watson AL, Freeman LM. Assessment of methods of evaluating sarcopenia in old dogs. *American Journal of Veterinary Research* 2012;73:1794–800.
16. Freeman L, Becvarova I, Cave N et al. WSAVA Nutritional Assessment Guidelines. *Journal of Small Animal Practice* 2011;52:385–396.
17. Michel KE, Anderson W, Cupp C, Laflamme DP. Correlation of a feline muscle mass score with body composition determined by dual-energy x-ray absorptiometry. *British Journal of Nutrition* 2011;10:S57–S59.
18. AAFCO. *2018 Official Publication of the Association of American Feed Control Officials*. 2018.
19. United States Department of Agriculture Agricultural Research Service. *USDA Food Composition Database*. Available from: https://ndb.nal.usda. gov/ndb/.

20. German AJ, Holden SL, Gernon LJ, Morris PJ, Biourge V. Do feeding practices of obese dogs, before weight loss, affect the success of weight management? *British Journal of Nutrition* 2011;10:S97–S100.
21. Wakshlag JJ, Struble AM, Warren BS et al. Evaluation of dietary energy intake and physical activity in dogs undergoing a controlled weight-loss program. *Journal of the American Veterinary Medical Association* 2012;24:413–419.
22. Flanagan J, Bissot T, Hours M-A, Moreno B, Feugier A, German AJ. Success of a weight loss plan for overweight dogs: The results of an international weight loss study. *PLOS ONE* 2017;12:e0184199.
23. Flanagan J, Bissot T, Hours M-A, Moreno B, German AJ. An international multi-centre cohort study of weight loss in overweight cats: Differences in outcome in different geographical locations. *PLOS ONE* 2018;13:e0200414.
24. Linder DE, Freeman LM, Morris P et al. Theoretical evaluation of risk for nutritional deficiency with caloric restriction in dogs. *Veterinary Quarterly* 2012;32:123–129.
25. Linder DE, Freeman LM, Holden SL, Biourge V, German AJ. Status of selected nutrients in obese dogs undergoing caloric restriction. *BMC Veterinary Research* 2013;9:19.
26. Bissot T, Servet E, Vidal S et al. Novel dietary strategies can improve the outcome of weight loss programmes in obese client-owned cats. *Journal of Feline Medicine and Surgery* 2010;12:104–112.
27. Levine ED, Erb HN, Schoenherr B, Houpt KA. Owner's perception of changes in behaviors associated with dieting in fat cats. *Journal of Veterinary Behavior: Clinical Applications and Research* 2016;11:37–41.
28. Westerterp-Plantenga MS, Nieuwenhuizen A, Tomé D, Soenen S, Westerterp KR. Dietary protein, weight loss, and weight maintenance. *Annual Review of Nutrition* 2009;29:21–41.
29. Westerterp-Plantenga MS, Lemmens SG, Westerterp KR. Dietary protein: Its role in satiety, energetics, weight loss and health. *British Journal of Nutrition* 2012;10:S105–S112.
30. Blanchard G, Nguyen P, Gayet C, Leriche I, Siliart B, Paragon B-M. Rapid weight loss with a high-protein low-energy diet allows the recovery of ideal body composition and insulin sensitivity in obese dogs. *Journal of Nutrition* 2004;13:2148S–2150S.
31. Hannah SS, Laflamme DP. Increased dietary protein spares lean body mass during weight loss in dogs [Abstract]. *Journal of Veterinary Internal Medicine* 1998;12:224.
32. National Research Council of the National Academies. Nutrient requirements and dietary nutrient concentrations. In: *Nutrient Requirements of Dogs and Cats*. Washington, DC: National Academies Press; 2006, pp. 354–370.
33. Laflamme DP, Hannah SS. Increased dietary protein promotes fat loss and reduces loss of lean body mass during weight loss in cats. *The International Journal of Applied Research in Veterinary Medicine* 2005;5:2–68.
34. Jackson JR, Laflamme DP, Owens SF. Effects of dietary fiber content on satiety in dogs. *Veterinary Clinical Nutrition* 1997;4:130–134.

35. Jewell DE, Toll PW. Effects of fiber on food intake in dogs. *Veterinary Clinical Nutrition* 1996;3:115–118.
36. Butterwick RF, Markwell PJ. Effect of amount and type of dietary fiber on food intake in energy-restricted dogs. *American Journal of Veterinary Research* 1997;58:272–276.
37. Heuberger R, Wakshlag J. The relationship of feeding patterns and obesity in dogs. *Journal of Animal Physiology and Animal Nutrition* 2011;95:98–105.
38. Loureiro BA, Sakomura NK, Vasconcellos RS et al. Insoluble fibres, satiety and food intake in cats fed kibble diets. *Journal of Animal Physiology and Animal Nutrition* 2017;10:824–834.
39. Kanarek RB. Availability and caloric density of the diet as determinants of meal patterns in cats. *Physiology & Behavior* 1975;15:611–618.
40. Prola L, Dobenecker B, Kienzle E. Interaction between dietary cellulose content and food intake in cats. *Journal of Nutrition* 2006;13:1988S–1990S.
41. Hirsch E, Dubose C, Jacobs HL. Dietary control of food intake in cats. *Physiology & Behavior* 1978;20:287–295.
42. Weber M, Bissot T, Servet E, Sergheraert R, Biourge V, German AJ. A high-protein, high-fiber diet designed for weight loss improves satiety in dogs. *Journal of Veterinary Internal Medicine* 2007;21:1203–1208.
43. German AJ, Holden SL, Bissot T, Morris PJ, Biourge V. A high protein high fibre diet improves weight loss in obese dogs. *Veterinary Journal* 2010;18:294–297.
44. Bennett N, Greco DS, Peterson ME, Kirk C, Mathes M, Fettman MJ. Comparison of a low carbohydrate-low fiber diet and a moderate carbohydrate-high fiber diet in the management of feline diabetes mellitus. *Journal of Feline Medicine and Surgery* 2006;8:73–84.
45. Hoenig M, Thomaseth K, Waldron M, Ferguson DC. Insulin sensitivity, fat distribution, and adipocytokine response to different diets in lean and obese cats before and after weight loss. *American Journal of Physiology: Regulatory, Integrative and Comparative Physiology* 2007;29:R227–R234.
46. Wei A, Fascetti AJ, Villaverde C, Wong RK, Ramsey JJ. Effect of water content in a canned food on voluntary food intake and body weight in cats. *American Journal of Veterinary Research* 2011;72:918–923.
47. Michel KE, Bader A, Shofer FS, Barbera C, Oakley DA, Giger U. Impact of time-limited feeding and dietary carbohydrate content on weight loss in group-housed cats. *Journal of Feline Medicine and Surgery* 2005;7:49–55.
48. Cline M, Witzel A, Moyers T, Kirk C. Comparison of high fiber and low carbohydrate diet on owner-perceived satiety of cats during weight loss. *Am J Anim Vet Sci* 2012;7:18–25.
49. German AJ, Holden SL, Mason SL et al. Imprecision when using measuring cups to weigh out extruded dry kibbled food. *Journal of Animal Physiology and Animal Nutrition* 2011;95:368–373.

50. Murphy M, Lusby AL, Bartges JW, Kirk CA. Size of food bowl and scoop affects amount of food owners feed their dogs. *Journal of Animal Physiology and Animal Nutrition* 2012;96:237–241.
51. German AJ, Holden SL, Mather NJ, Morris PJ, Biourge V. Low- maintenance energy requirements of obese dogs after weight loss. *British Journal of Nutrition* 2011;10:S93–S96.
52. Laflamme D, Kuhlman G. The effect of weight-loss regimen on subsequent weight maintenance in dogs. *Nutrition Research* 1995;15:1019–1028.
53. Nagaoka D, Mitsuhashi Y, Angell R, Bigley KE, Bauer JE. Re-induction of obese body weight occurs more rapidly and at lower caloric intake in beagles. *Journal of Animal Physiology and Animal Nutrition* 2010;94:287–292.
54. German AJ, Holden SL, Morris PJ, Biourge V. Long-term follow-up after weight management in obese dogs: The role of diet in preventing regain. *Veterinary Journal* 2012;19:65–70.
55. Deagle G, Holden SL, Biourge V, Morris PJ, German AJ. Long-term follow-up after weight management in obese cats. *Journal of Nutritional Science* 2014;3:e25.

6 肥胖症宠物的行为管理策略

6.1 引言

宠物肥胖是一种复杂的营养失调症,通常需要全面的管理,除了执行标准饮食和运动计划外,还需要了解人和动物之间的相互影响,才能达到成功减肥的目的。与宠物主人进行有效的沟通,可以帮助兽医团队了解每个家庭与宠物之间的独特关系,并在体重管理过程中保持人与动物之间的良好情感纽带。因此,宠物肥胖的治疗可以包含多种模式,包括从医学和心理学双重角度入手,从而成功实现对减肥的行为管理。虽然肥胖可能是一种具有挑战性的疾病,伴随宠物的一生,但兽医可以不拘泥于标准治疗方案,还可将宠物主人与宠物之间独特的关系纳入减肥计划。这不仅加强了宠物主人与宠物之间的联系,还能提高治疗肥胖症的成功率。

6.2 宠物主人主观认知对宠物行为和肥胖症的影响

在世界范围内对犬猫的研究结果表明,导致宠物肥胖的危险因素在不同的研究中是不同的[1-8]。在一项流行病学研究中,与宠物主人相关的因素几乎是导致宠物肥胖的共性危险因素[3]。这些因素包括宠物主人的年龄,以及由主人控制的诸如进食频率和运动量等行为。许多与宠物主人有关并导致宠物肥胖的风险因素,提示兽医需要充分了解宠物和主人的行为、家庭动态,以及宠物与主人的关系是如何影响宠物体重并影响减肥效果的(表6.1)。

表 6.1 与宠物主人相关的导致宠物肥胖的危险因素

危险因素
饲喂剩饭
提供零食点心
宠物主人年龄增大
宠物主人收入减少
宠物犬运动量减少
宠物主人对宠物身体状况的熟悉程度

来源:Courcier EA et al., *Journal of Small Animal Practice* 2010;51:362-367; Rohlf VI et al., *Journal of Applied Animal Welfare Science* 2010;13:213-236; White GA et al., *Journal of Small Animal Practice* 2011;52:622-626; Bland IM et al., *Preventative Veterinary Medicine* 2009;92:333-340.

6.2.1 对宠物肥胖症的认知

宠物主人低估了宠物的体况评分是导致宠物肥胖的一个重要危险因素。在2项独立的研究中,即使得到了由兽医或训练有素的评估师对其宠物体况评分的评估结果,仍有39%的宠物主人低估了宠物的体况评分[9,10]。尽管Bland和他的同事们也强调了宠物主人们对理想体重存在误解[11],然而在宠物主人得知宠物的"准确"体况评分分数高于理想体重的情况下,仍有一半宠物主人不愿意承认他们的宠物超重。图6.1显示,宠物医院把体况评分表和体重秤放在一起对比,试图直观上缩小这种认知差异。宠物主人和兽医对宠物理想身体状况的评判差异可能源于公众普遍认为宠物超重才美这样的标准。例如,一项关于表演犬的研究显示,几乎每5只表演犬中就有1只超重[12],同样,表演猫中也出现这种情况[13]。宠物主人的认知会影响他们的行为以及他们对宠物体重管理计划能否坚持下去。例如,有研究表明,在考虑喂食行为时,食物碗和食物勺的大小会影响宠物主人给爱宠的喂食量[14]。此外,即使是兽医工作人员对标准喂食量的估计也不准确,从低估18%到甚至高估80%不等[15]。由于存在许多与宠物主人行为有关的,导致宠物肥胖的风险因素,加上宠物主人对宠物身体状况和喂养方法的错误认识,因此,宠物的体重管理除了执行标准饮食和运动治疗方案外,还应规范宠物主人及其宠物的行为。

6.2.2 了解宠物主人的想法

当与宠物主人讨论宠物肥胖问题时,要收集信息以了解宠物完整的饮食史,花时间充分了解宠物和宠物主人的生活环境和行为将有助于评估宠物主是否愿意做出改变,以及是否有能力执行宠物的体重管理计划。有些宠物主人可能没有下决心为了宠物减肥而改变自己的行为,仍然在犹豫是否对其宠物进行体重管理。在这种情况下,客户教育和监督管理可能是最好的办法,然后再要求宠物主彻底改变观念,实施全面的减肥计划。然而,对于那些已经准备好或已经开始为宠物的体重管理采取行动的宠物主人,他们应该从具体的个性化方案和反馈中受益。例如,宠物主可能认为他们只需要为宠物选择一种"减肥餐",但不确定选择哪一种,或者没有意识到他们可能需要改变的是自己为宠物提供除主食以外食品的这种行为习惯。提供模糊不清的指导是没有效果的,如"改吃清淡的食物"或"少吃一点犬粮",这些指导建议会使宠物主人更加困惑,而且也并没有说清楚是什么样的行为导致了肥胖,例如,提供过量食物导致宠物摄入过高热量从而导致了肥胖[16]。

让宠物主人参与,了解他们对宠物体重相关问题以及营养需要的看法,这可以让兽医了解宠物主人是否愿意改变,所有这些信息都可能会影响兽医之前给出的减肥方案和宠物主人行为改变的建议。关于评估宠物主人是否已准备就绪的指

图 6.1　一家宠物医院将宠物体况评分表与体重秤结合起来，鼓励人们谈论宠物理想的身体状况

南，在本书的其他地方已进行过描述[17]。虽然与宠物主讨论宠物体重问题是一项很有挑战性的事情，但是了解宠物主的观点可以指导管理，在以下这些讨论中描述了辅助兽医健康管理团队与宠物主谈话的技巧[18]。

通过这些讨论，宠物主人可能表示他们有兴趣执行兽医建议的任何计划，或者可能解释他们的能力有限，这些信息均可能指导体重管理方案和行为规范。对于那些对任何类型的计划都感兴趣的宠物主人,兽医可以只关注医学因素指导体重管理,如根据肥胖宠物的营养或能量需求选择饮食。然而,在其他情况下,兽医可能需要平衡医疗、社会或心理因素来制定一个宠物体重管理计划。例如,在饮食选择和肥胖治疗中,可能要考虑宠物主人的工作安排和喂食习惯。如果宠物主人由于工作安排每天只能有2次遛犬时间,由于高纤维的食物会产生更多的粪便,那么这种安排显然是不适合的。在这种情况下,虽然纤维可能会帮助一些宠物有饱腹感,但如果不考虑喂食行为而向宠物主人推荐这种食物,宠物主人可能会因为宠物在家里排便而放弃该计划,并认为体重管理困难重重。家庭喂养的其他宠物也同样要实施相同的喂养方法以保持一致。通过了解宠物主人的背景和想法,兽医可以通过宠物主人能够接受和参与的方式进行体重管理,尤其是行为的改变,以获得最大的成功。

6.3 将宠物主人心理和行为的管理纳入宠物肥胖症治疗方案

成功管理宠物体重的一个重要方面是处理好人和宠物的关系,这可以在肥胖症的预防和治疗中发挥积极作用。宠物主在执行体重管理计划时,除了医疗支持外,可能还需要社会和心理支持。了解人与动物关系的动态变化,有助于兽医与客户制定成功的治疗方案。本节将讨论把社会科学理论应用于实践的过程,并将对体重管理项目中社会心理方面的实际管理案例纳入标准营养管理体系中。

6.3.1 与宠物主人讨论宠物的肥胖问题

不仅仅是兽医从业者们不喜欢探讨宠物肥胖的问题,许多从事人类医疗保健服务的人员同样不情愿与人交流体重管理问题[19]。与宠物肥胖症相似,人类肥胖及其相关合并症也是一个严重的公共健康问题,在过去的30年中,人类肥胖的发病率已增加了2倍[20]。与兽医学类似,制定有效的和可持续的干预措施来治疗肥胖仍然是人类医学的一个首要任务和挑战。由于每个肥胖患者或肥胖宠物可能需要的是针对他们的个性化治疗方案,这也使得人类医疗健康服务人员和兽医面临着更多的挑战。在兽医学中,体重管理时将宠物主人行为以及他与宠物的关系作为首先考虑的因素是很有帮助的。这里包括3个方案,在其他的地方已经详细讨论过[18]。

6.3.1.1 基于信息交流的案例

根据作者的经验,一些宠物主人非常困惑,因为之前没有人指出他的宠物体重存在潜在的健康风险。由于缺乏专业的建议,宠物主人误认为其宠物处于理想或者健康的体重。您如果不信,可以问问之前是否有人曾经讨论过宠物体重。这里也包括对肥胖的合并症,以及对健康后续影响的讨论,例如超重猫有易患糖尿病的风险。建议可以不经意的开启谈话,以一种中立的不带有煽动、指责和情绪化的方式谈及宠物的健康信息,这种顺其自然的谈话能获得宠物主人良好的回应。

6.3.1.2　与人类进行比较的案例

那些自身也超重的宠物主人们,一旦谈及宠物肥胖的话题,常常还会提出他们在肥胖问题上遇到的类似困难。虽然兽医为人类提供体重管理建议不符合兽医职业行为准则,但借鉴相似之处可能有助于探讨宠物肥胖问题。例如,在一篇综述文章中作者提到的一个例子可以提供借鉴,当宠物主人说,"我自己都宁愿吃饼干也不吃粗粮",可以尝试着这样回应,"适度的零食是可以的,但是,和我们人类一样,如果我们的宠物也一直只吃糖果或饼干,它们可能就会生病。您愿意为爱宠讨论制定出一个合理的体重管理计划吗?包括均衡的饮食结构和宠物喜欢的零食"[18]。

6.3.1.3　基于人类情感的案例

一些宠物主对他们的宠物有着强烈的情感依恋,从他们对待宠物的行为就反映了他们之间的紧密关系。例如,超过一半的宠物主人允许他们的宠物在自己的床上睡觉,还会做出一些将宠物拟人化的行为,如给宠物买礼物、做饭、穿衣服等[21]。这种情感可能会成为减肥的障碍,例如,有的客户说,"让我的宠物挨饿而不给它想要的东西是很残忍的"。面对这些情感陈述,可以用一种积极的方式来解释减肥对宠物的影响。例如,交谈开始就阐述减肥对宠物没有危害,也不会给宠物带来痛苦。而恰恰相反的是超重的宠物更容易感到疼痛,生活质量也更差,比如其活力会下降[22]。减肥不会伤害他们,却会让他们变得更好。此外,宠物"乞食"行为可能只是被训练成寻找宠物主人的方式,因为主人有食物。可以引入食物分配玩具或自动喂食器,以减少觅食这种行为,如图 6.2 所示。

图 6.2　一种自动喂食器,该自动喂食器可以设置,给宠物定时、定量的提供食物(由 PetSafe[R] 提供的 5 格宠物自动喂食器)

6.3.2 在最初的体重讨论中参考兽医的意见

在作者的临床经验中,情绪化的例子是诊室中非常常见的情况,有时可能是最具挑战性的事情,因为一些客户有强烈的固有信念和先前的体重管理理念。有研究表明,宠物主人对失去宠物的内疚或恐惧是造成他们不愿意参加减肥计划的主要原因[11]。然而,当前的数据表明,这种想法可能是一种误解,应该与宠物主进行探讨,这种探讨可能会促进宠物主人积极改变自己的行为而参与到减肥计划中[22]。例如,一项研究表明,宠物体重超标会降低生活质量,而在减肥成功后,这些指标(如活力和疼痛)会相应得到改善[22]。讨论肥胖对生活质量的积极影响,以及减肥所能带来的好处,这将有助于减轻宠物主因错误理解而产生的内疚感,进而可能对是否减肥犹豫不决。此外,虽然难以做到使宠物主将目前的肥胖症状与将来可能患某种疾病联系起来,但兽医可以强调肥胖可能对宠物目前的生活质量产生不利影响,从而促使其尽快改变行为。可以提供给宠物主一份经过验证的生活质量调查问卷,客观地衡量这些影响,以及随着时间的推移,体重管理是如何使之发生变化的[22]。在启动和实施体重管理计划时,兽医和宠物主讨论的共同目标,即宠物的健康和舒适,可以帮助宠物主轻松地进入交谈,并解决导致肥胖行为的根源。

6.4 通过解决情绪行为来制定有效的体重管理计划

拥有一只宠物以及宠物和主人之间的亲密关系可能对人类心理健康有益[23]。与动物简单互动的好处是能产生共情[24]、增加情感支持[25]、减少孤独感[26]和压力[27]。主人与宠物之间的关系如此的密切,他们的心理会受到积极的影响,所以对影响这一关系的体重管理计划必须进行调整,以保持这一良好关系,从而确保宠物主能够坚持下去并成功减肥。了解宠物主的行为以及对宠物的依恋有助于在兽医和宠物主之间建立信任感。在这种信任感的基础上,兽医和宠物主可以一起制定体重管理计划以加强主人与宠物的关系,而不会威胁到这种有益的亲密关系。

有趣的是,强烈的依恋感可能会影响宠物主人改变行为的意愿,从而对治疗产生积极或消极的影响。例如,与宠物关系密切的主人可能更愿意投入财力和时间来为自己的宠物进行兽医护理和寻求体重管理计划。然而,高度依恋也可能与情感依赖和导致肥胖的行为有关,比如一些有临床经验的宠物主会提供过量的食物(热量)来表达对宠物的喜爱之情。与动物有强烈情感关系的主人可能不太愿意改变喂食行为,比如改变其提供高热量食物的习惯,因为这些行为习惯某种程度上代表了他们的亲密关系。例如,在询问宠物主的饮食史时,让他们描述一下他们与宠物的日常生活可能会有帮助,这将提供开放式的答案。进一步地提问,比如,

"您强烈希望我们的减肥计划中包括哪些活动?"这会让宠物主人觉得维持他们与宠物的关系是最重要的,如果不以某种方式加以配合,可能会破坏减肥计划。

根据作者的经验,这些问题往往揭示出一些不可商榷的因素,这些因素可以结合起来解决,以增加对体重管理计划坚持的信念。例如,如果宠物主人说每天晚上主人看电视的时候宠物需要吃些零食,那么可以选择低热量的食物来替代,以便在坚持体重管理计划的同时保持他们的关系。通过这种方式,宠物主仍然可以按照自己的意愿去做(提供零食并和宠物分享时光),但是食物本身或食物中的热量需要改变,这不会对他们的关系产生不良影响。在另一种情况下,如果宠物主人认为,与不喜欢体育活动的宠物一起体育运动可能在情感上具有挑战性,那么可以将喂食时间与运动结合起来,让宠物和宠物主人都能从中获得快乐。例如,不把食物放在碗里,而是将干粮分放在整个房子里,也可以设计猫或犬的取物游戏。在这种情况下,了解宠物主和宠物之间的情感可以帮助兽医和宠物主找到应对方法从而成功减肥。

6.5 考虑家庭成员互动在管理肥胖行为中的作用

了解不同家庭环境下的人与动物关系,可以帮助兽医根据家庭情况制定可行的减肥计划,并在现有家庭动态中实施可行的行为改变。在收集宠物饮食习惯和讨论可能的减肥计划的同时,也应该了解家庭其他成员和他们与宠物之间的关系。有必要询问客户一些非常具体的问题,如哪些家庭成员负责宠物护理的方方面面,如喂养、运动和梳洗等(表6.2)。这些信息将指导制定适合现有家庭结构的体重管理计划。此外,讨论人与动物的家庭关系可以帮助我们提前解决体重管理计划执行过程中可能遇到的障碍。例如,家庭中有好几个人,只有在了解了整个家庭成员如何喂养宠物后才能获得准确的饮食记录。根据作者的经验,家庭成员可能不知道家里其他人为宠物提供食物的详细情况。有时,这些喂食行为是故意隐瞒家里的其他人,也有时候,宠物主人不可能针对宠物每一次食用的食物进行沟通。

表6.2 饮食史调查中应包含的相关问题示例

问题示例
宠物可以吃其他宠物的食物吗?
家里住多少人?
家里都有谁来喂您的宠物?
您的家人如何决定给您的宠物喂多少?
描述一下宠物的用餐时间。
您的宠物会讨食吗?如果是,请描述具体行为。

6 肥胖症宠物的行为管理策略

在上述 2 种情况下,要解决多个主人的喂养方式不同的问题,有效的方法是在所有家庭成员都在场的情况下,讨论宠物每日的饮食安排。这可以包括每天所需的完整和均衡的食物,以及在规定热量内的各种零食。为了确保宠物不会被无意识地喂了好几次,或者喂得比规定的量多,家庭成员们可以在每天晚上准备好宠物第 2 天的食物盒,这些食物加起来可满足所需要的能量,包括每顿餐食和各种各样的零食。在这种情况下,家里的每个人都可以亲眼看到,他们的宠物在一天中确实得到了很多的食物,而食物只是被许多家庭成员分散的分配了。家人们共同制作一个宠物食物盒,也可以增加家庭成员之间的交流,而且也可使家庭成员和宠物之间的关系更加紧密。满足家庭中所有成员的需求可以最大限度地坚持体重管理计划,同时增加实施体重管理计划的成功率并提高乐趣。

6.6 宠物与主人共同减肥的互动行为

虽然大多数行为管理的重点是要调整行为习惯,使之有益于宠物的健康和体重状况,其实人与动物之间的亲密关系同样也有益于人类的身体健康。一项研究结果表明[28],参与宠物的减肥计划对宠物主的身体健康和运动水平有积极的影响[29]。此外,与正常体重宠物的主人相比,体重超标宠物的主人可能对他们的宠物依恋程度更高,而从同伴那里获得的社会支持却较少[29]。如果超重宠物的主人确实在向他们的宠物而不是社会中同龄人寻求社会支持,那么让宠物主人参与一些对他们自己和宠物都有益的行为中来会特别有帮助,并增加宠物减肥成功的机会。然而,不管体重状况如何,所有的宠物主都可以通过健康的行为来提升他们与宠物的关系,比如锻炼,而不是像一些主人那样只是通过为宠物提供额外热量的零食来表达他们的爱。图 6.3 就显示了

图 6.3 宠物主人可以与犬一起享受散步和游泳等体育运动的时光

宠物和主人共同参加体育活动的机会（例如，散步或游泳等）。

综合治疗方案应以人类和动物健康为目标。以家庭为导向的肥胖治疗项目，可以让体重超重的家庭成员和他们的宠物一起变得更健康，这也可能对兽医和人类医学中的健康问题产生巨大的影响。在针对这类计划最有效的方法开展进一步的研究之前，兽医应该更专注于保持有益的、健康的人与动物的关系。最好的办法是了解主人和宠物之间的关系，培养已经存在的有益健康的行为，在不牺牲人与动物之间有益关系和纽带的情况下，开发一些新的行为习惯，替代那些易于使人和宠物肥胖的行为习惯。

6.7 结论

许多风险因素，再加上人类与宠物关系的复杂性，可能会营造出一个导致宠物肥胖的生活环境。从积极的方面来说，将肥胖症的许多医学和社会科学方面结合起来考虑，可以提高体重管理的成功率。虽然肥胖症的治疗需要密集而全面的管理，但肥胖症治疗还有很多新的方面，甚至还有许多方面还有待研究，但这可以给主人、宠物和兽医带来有益和愉快的经历。采取全面的体重管理计划包括了解和整合主人与宠物之间的复杂关系，虽然这超出了标准的营养管理范畴，却可以提高减肥的成功率。

参考文献

1. McGreevy PD, Thomson PC, Pride C, Fawcett A, Grassi T, Jones B. Prevalence of obesity in dogs examined by Australian veterinary practices and the risk factors involved. *Veterinary Record* 2005;156:695–702.
2. Lund EM, Armstrong PJ, Kirk CA, Klausner JS. Prevalence and risk factors for obesity in adult dogs from private US veterinary practices. *International Journal of Applied Research in Veterinary Medicine* 2006;4:177–186.
3. Courcier EA, Thomson RM, Mellor DJ, Yam PS. An epidemiological study of environmental factors associated with canine obesity. *Journal of Small Animal Practice* 2010;51:362–367.
4. Scarlett JM, Donoghue S, Saidla J, Wills J. Overweight cats: Prevalence and risk factors. *International Journal of Obesity and Related Metabolic Disorders* 1994;18:S22–S28.
5. Lund EM, Armstrong PJ, Kirk CA, Klausner JS. Prevalence and risk factors for obesity in adult cats from private US veterinary practices. *International Journal of Applied Research in Veterinary Medicine* 2005;3:88–96.
6. Colliard L, Paragon B-M, Lemuet B, Bénet J-J, Blanchard G. Prevalence and risk factors of obesity in an urban population of healthy cats. *Journal of Feline Medicine and Surgery* 2009;11:35–40.

7. Courcier EA, O'Higgins R, Mellor DJ, Yam PS. Prevalence and risk factors for feline obesity in a first opinion practice in Glasgow, Scotland. *Journal of Feline Medicine and Surgery* 2010;12:746–753.
8. Cave NJ, Allan FJ, Schokkenbroek SL, Metekohy CA, Pfeiffer DU. A cross-sectional study to compare changes in the prevalence and risk factors for feline obesity between 1993 and 2007 in New Zealand Preventative. *Veterinary Medicine* 2012;107:121–133.
9. Rohlf VI, Toukhsati S, Coleman GJ, Bennett PC. Dog obesity: Can dog caregivers' (owners') feeding and exercise intentions and behaviors be predicted from attitudes? *Journal of Applied Animal Welfare Science* 2010;13:213–236.
10. White GA, Hobson-West P, Cobb K, Craigon J, Hammond R, Millar KM. Canine obesity: Is there a difference between veterinarian and owner perception? *Journal of Small Animal Practice* 2011;52:622–626.
11. Bland IM, Guthrie-Jones A, Taylor RD, Hill J. Dog obesity: Owner attitudes and behavior. *Preventative Veterinary Medicine* 2009;92:333–340.
12. Corbee RJ. Obesity in show dogs. *Journal of Animal Physiology and Animal Nutrition* 2013;97:904–910.
13. Corbee RJ. Obesity in show cats. *Journal of Animal Physiology and Animal Nutrition* 2014;98:1075–1080.
14. Murphy M, Lusby AL, Bartges JW, Kirk CA. Size of food bowl and scoop affects amount of food owners feed their dogs. *Journal of Animal Physiology and Animal Nutrition* 2012;96:237–241.
15. German AJ, Holden SL, Mason SL, Bryner C, Bouldoires C, Morris PJ, Deboise M, Biourge V. Imprecision when using measuring cups to weigh out extruded dry kibbled food. *Journal of Animal Physiology and Animal Nutrition* 2011;95:368–373.
16. Linder DE, Freeman LM. Evaluation of calorie density and feeding directions for commercially available diets designed for weight loss in cats and dogs. *Journal of the American Veterinary Medical Association* 2010;236:74–77.
17. Churchill J. Increase the success of weight loss programs by creating an environment for change. *Compendium: Continuing Education for Veterinarians* 2010;32:E1–E4.
18. Linder D, Mueller M. Pet obesity management: Beyond nutrition. *Veterinary Clinics of North America: Small Animal Practice* 2014;44:789–806.
19. Phillips K, Wood F, Kinnersley P. Tackling obesity: The challenge of obesity management for practice nurses in primary care. *Family Practice* 2014;31:51–59.
20. Ogden CL, Carroll MD, Kit BK, Flegal KM. Prevalence of obesity in the United States, 2011–2012. *Journal of the American Medical Association* 2014;311:806–814.
21. Krahn LE, Tovar MD, Miller B. Are pets in the bedroom a problem? *Mayo Clinic Proceedings* 2015;90:1663–1665.
22. German AJ, Holden SL, Wiseman-Orr ML, Reid J, Nolan AM, Biourge V, Morris PJ, Scott EM. Quality of life is reduced in obese dogs but improves after successful weight loss. *Veterinary Journal* 2012;19:428–434.

23. Budge RC, Spicer J, Jones B, St. George R. Health correlates of compatibility and attachment in human-companion animal relationships. *Society and Animals* 1998;6:219–234.
24. Melson GF, Peet S, Sparks C. Children's attachment to their pets: Links to socio-emotional development. *Children's Environments Quarterly* 1992;80:55–65.
25. Kurdek LA. Pet dogs as attachment figures for adult owners. *Journal of Family Psychology* 2009;23:439–446.
26. Stanley IH, Yeates Conwell BA, Bowen C, Van Orden KA. Pet ownership may attenuate loneliness among older adult primary care patients who live alone. *Aging and Mental Health* 2014;18:394–399.
27. Barker SB, Dawson KS. The effects of animal-assisted therapy on anxiety ratings of hospitalized psychiatric patients. *Psychiatric Services* 1998;49:797–801.
28. Kushner RF, Blatner DJ, Jewell DE, Rudloff K. The PPET study: People and pets exercising together. *Obesity (Silver Spring)* 2006;14:1762–1770.
29. Stephens MB, Wilson CC, Goodie JL, Netting FE, Olsen CH, Byers CG. Health perceptions and levels of attachment: Owners and pets exercising together. *Journal of the American Board of Family Medicine* 2012;25:923–926.

7 运动在宠物肥胖症治疗中的作用

7.1 引言

运动在犬猫肥胖症治疗中的作用尚不完全清楚。但基于对危险因素的多项流行病学研究,发现缺乏运动容易导致犬类肥胖[1,2]。许多兽医会给宠物主提供一些建议,告诉他们让超重或肥胖的宠物增加运动有助于减肥。然而,在接受能量控制基础上加入适当运动,这种方案并未在科学文献中进行严格的评价。人类现有的证据表明,在减肥过程中,体育活动可能有治疗效果,但这些通常与心血管健康和其他继发性疾病的改善有关,而不是直接影响能量摄入或直接促进减肥方案的成功[3-6]。运动对犬和猫也有类似的益处,尤其是当肥胖的动物患有骨关节炎、运动障碍和肌肉萎缩等其他的合并症时。对于体重正常的犬,运动对热量消耗的影响与运动距离有关[7],对超重的动物同样如此,但这样的犬和猫可能对运动有最大的耐受性,也降低了运动在消耗能量中的贡献率。

7.2 运动与肥胖症风险

运动量的减少是导致犬类肥胖的一个危险因素。一项研究发现,每周运动1 h,肥胖风险率降低4%[1]。一项对中国宠物犬的研究发现,能够自由活动的犬不容易肥胖(OR = 0.685)[8]。运动强度的大小与降低肥胖风险的幅度往往关系不大。例如,澳大利亚的肥胖犬平均每周运动2.5 h,而非肥胖的犬每周运动3.4 h[2]。经常运动的犬可能以更快的速度运动。运动时跑步的犬比运动时走路的犬活跃度高出50%。因此,这些犬将走更远的距离,消耗更多的能量。无论进行哪种类型的运动,每小时都会降低肥胖风险90%[2]。其他研究表明,动物一天内的活动量与肥胖风险成反比[9],尽管没有直接证据,猫也可能表现出类似的趋势。与体重正常猫相比,超重猫的主人与猫的互动可能更少[10]。遗憾的是,这些研究只建立了相关性,而不是因果关系,因此还需谨慎诠释。很可能是这样的,犬因不经常运动或者因为长期运动不耐受引起了继发性肥胖或者并发损伤而引起了继发性肥胖。例如犬,肥胖症和前十字韧带断裂之间存在相关性[11],这可能是肥胖症或是后遗症导致的,但仍会影响活动。因此,导致肥胖的因素可能主要是饮食习惯、品种、合并症和主人因素,而不是缺乏运动 [1,11]。

如果证实了活动减少和随后体重增加之间存在因果关系,那么宠物主的生活习惯和对宠物的态度会影响动物的体育运动并可能引起肥胖风险。预测可能影响宠物犬运动量的因素包括:单犬家庭、中型犬和大型犬、女性主人和人口较少的家庭 [12,13]。肥胖犬的主人比正常犬的主人更不喜欢运动 [14],而在另一项研究中,宠物主人的态度与宠物身体状况得分无关 [15]。现有的证据表明,宠物主人对宠物的态度会影响宠物的运动量,但运动量减少不一定会导致肥胖,因为这可能是多种因素造成的。

7.3 运动治疗人类肥胖症

运动与人类肥胖症管理之间的联系比犬或猫的研究更为广泛深入。然而,由于物种的差异,对人类文献研究结果的解读是否适用于宠物应该保持更谨慎的态度。这对犬来说是正确的,因为犬具有与人类相似的饮食需求,而对于专属食肉动物猫来说更是可信度较高。据推测,运动可以增加肥胖者的脂肪氧化,增加与运动强度相对应的数小时代谢率,并影响肌肉分泌型细胞因子反应的类型,这可能会抵消一些脂肪因子 [16-18]。然而,犬的脂肪基础氧化水平比人高 [7,19,20]。这2个物种在肥胖期间显示出脂肪因子水平有重要差异,在体重减轻后并没有在犬体内观察到总脂联素和高分子量脂联素的减少,这与人类实验中观察到的结果不同 [21]。针对猫的运动生理学研究资料极少,这就很难与人类研究文献资料做对比。

可以通过运动来减少肥胖对心血管的影响,而运动对减轻体重的实际效果是很有限的 [6]。研究表明,长距离运动(每周超过20英里)会导致人体每月体重下降不到0.4%,也可能是因为这些运动建议执行不到位 [6,22]。对于人类来说,适度的运动可以减少体重反弹,并有助于人们长期保持体重。每周保持3 h或更长时间的运动就可以达到这种效果,这个数据是接近已有记录的一些肥胖犬活动量的最低水平 [2,6]。相比于那些仅随访12个月或36个月之后才计算的案例,人类在控制饮食又增加运动的减肥方式中,往往在18个月时达到最佳的减肥效率,这说明这种减肥方式的最大效益可能存在明显的延迟,并在某一时间点达到峰值 [23]。个体对运动反应的异质性和达到临床效果所需的大量运动反映的是临床建议,即超重个体期望通过此类干预获得少量的总体重减轻(<2%的起始体重)。然而,中等强度的有氧运动和饮食相结合,而不是单纯的控制饮食,可以改善心血管健康,改善肥胖对健康带来的一些不良后果,如心脏功能下降、血压升高、胰岛素抵抗、糖尿病和血脂异常等 [3-5]。这些益处也可能会在犬和猫身上体现出来,但这些研究结果

因物种差异导致适用性有局限性,详情还需要进一步的科学研究。例如,猫和人类的肥胖症与非胰岛素依赖型糖尿病有关,犬不是这种情况。然而,还未发现运动在降低猫糖尿病发病率和死亡率中的作用。

7.4 犬和猫合并症、运动和肥胖症

许多与犬或猫肥胖症有关的合并症,在本书中已有详细描述(第3章),并可能与推荐给肥胖宠物的运动方案有关。尽管数据有限,但仍表明肥胖会像影响人类心血管健康那样影响犬的心血管健康。有记录表明,犬的运动耐力会下降,同时左心房压会显著升高,血压会轻度升高[24]。某些肥胖的犬在呼吸急促时气道阻力发生了改变,这可能是观察到一些超重动物抗拒运动的一个原因[25]。越来越多的肥胖犬会出现骨关节炎的影像学和临床症状[26],但有趣的是,在实验中,与对照组相比,增加了负重并在跑步机上跑步的犬并没有发现关节炎加重[27]。这可能说明,肥胖对激素的影响会导致关节炎,或者动物在遗传上易患关节炎,关节负荷增加会在肥胖病情发展后恶化。另外,肥胖可能会减少活动,这可能是由于较少的运动量使得机械力降低,从而有利于软骨细胞健康和营养供应[28]。推荐使用定期的低强度运动用于治疗人类和犬的关节炎[29,30],减轻体重可以改善患骨关节炎动物的负重,减少全身炎症的标志物[21,31]。尚未在猫上开展大量的科学研究来关注各种运动方案的功能和效果。

7.5 运动在犬和猫肥胖症治疗中的应用

运动被认为有助于提高减肥计划的效果,但这主要是基于专家意见,而不是临床经验[32]。这些建议并不是来自科研人员前瞻性调查研究,但通常还是会建议宠物步行30 min左右,或者每天游泳15 min,每周游泳5~7次[10]。虽然体重管理中一直认为超重的猫和犬需要增加活动,但没有详细或是明确证据证明[33]。如果有的话,很少有研究能确定运动在犬类和猫科动物减肥方案中的最佳作用。

根据高级计步器的报告,与不活动的犬(42 kcal/MBW)(MBW, metabolic body weight,代谢体重)相比,身体活动较多的超重犬消耗的能量更多(54 kcal/MBW)[34]。每1 000步运动量与能量摄入增加1 kcal/kg$^{0.75}$呈轻度相关($r = 0.36$)。水下跑步机被纳入一项减肥项目中,该计划每周使超重的犬减重1.5%[35]。通过处方粮控制热量来减肥的宠物犬也同时接受了平均13次水疗法,平均行走距离为0.97 km。要求宠物主人们每天遛犬,但这并没有量化指标。作者认为,宠物减肥速度大于回顾性的评估结果,很遗憾,这种分析是对多种干预措施进行的回顾性评估。相比于宠

物主人在家里控制能量摄入的治疗方式,在家和诊所接受物理治疗的患有骨关节炎的超重犬,其体重减轻比前者多46%[36]。物理治疗主要是指在家里积极运动,如步行1h,但可惜的是,没有2组宠物的能量摄入情况对比数据。

最具代表性的由兽医指导的运动干预数据来自针对最近的一项减肥计划的研究,该研究针对一小群超重犬,进行单独的饮食管理或在饮食的基础上额外增加每周3次的水疗(30 min)和地面跑步机行走(30 min)[37]。爬坡、行走或小跑的速度以及水深都会根据治疗情况随之调整,犬的主人们得到的建议不仅仅是控制饮食,而是鼓励宠物主人在家通过运动来锻炼患病犬。在研究期间,2组间犬的能量摄入量相同。经过运动锻炼的犬在12周内每周平均减重1.16%,而饮食管理组每周平均减重1.08%,这在统计学上差异并不显著。2组均观察到静息心率的改善,尽管实验组治疗犬的加速计数增加,而节食犬组的加速计数减少,但2组间并无差异。运动疗法的益处,也是本实验最重要的发现显示,运动的犬瘦体重增加了4.6%,而久坐不动的犬瘦体重降低了2.5%。

在一项相关研究中,在登记在册的犬身上发现了相关转录组发生了有利的变化[38]。在研究期间,体重减少达到330g才能达到统计学差异。应用Wishnofsky法则[39],在运动36 h,大约19.7 km的距离后,将产生2 500 kcal的内源性能量消耗,约合计182 kcal/kg$^{0.75}$。由此得出结论,如果认为这些都是在诊所跑步机上运动的结果,那么计算得出每千米消耗能量9.2 kcal/(kg$^{0.75}$·km)。与基线相比,接受治疗的犬每周的加速计数增加了85万个,每百万计的能量值为245 kcal。有趣的是,在12周时,与基线相比,控制饮食的犬的加速度计计数减少了7.6%,这与另一项控制饮食的研究大致一致,该研究记录在同一时间点加速度计计数减少了11.3%[40]。

对于正常、健康的犬来说,各类运动的能量消耗如图7.1[7,41,42]所示。与躺下相比,坐着和站着分别需要多消耗30%和46%的能量,因此,犬活动保持多样性变化会影响每日的热量消耗[43]。犬散步时每走1 km大约需要能量1.2 kcal/kg$^{0.75}$。因此,按照一般建议,一只体型中等大小的犬在30 min的遛犬过程中只能增加1%的能量消耗[44]。犬跑步时所需能量约为每千米1.1 kcal/kg;与短腿犬相比,长腿的犬在跑相同距离时消耗的能量则更少[7,45],但大多数肥胖犬的体能都很差,跑不动。肘部高度的水下跑步机运动需要消耗1.9 kcal/(kg$^{0.75}$·km)[44]。给中型犬进行30 min的水下跑步机训练,只会增加1.5%的日常能量消耗[42]。关于动物游泳可以提高减肥成功率这种建议,目前还没有数据能用来评估它的效果。

在一项猫的强化训练中,使用了猫很努力才能够得着的喂食器、新玩具和攀爬结构,以及用食物刺激的运动项目,这些在与处方粮搭配使用时,就造成了有些运动量无法量化估计,因而与对照组相比,在统计学上有差异,但在临床上却表现得差异不明显[46]。如果将猫的光照时间延长到每天16 h,活动量增加了20%,但它

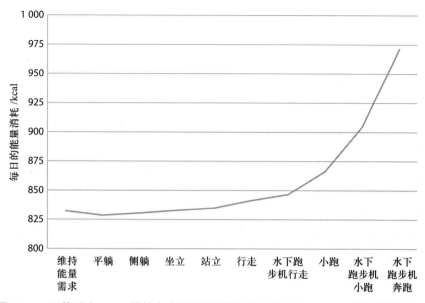

图 7.1　一只体重为 20 kg 的健康犬采取不同的运动形式活动 30 min，其能量消耗情况

的能量摄入量仅增加了 5%[47]。关于增加猫的喂食频率或在饮食中增加水分是否会增加猫的运动量，得到的结论不尽相同[48,49]。

尽管所有的运动建议都提供给了宠物主人，但运动还要取决于宠物主人的实际执行情况。除非宠物主坚定地实施该减肥计划，否则在减肥开始前存在的那些影响宠物运动的因素很可能仍然存在。来自宠物肥胖诊所提供的计步器数据表明，尽管建议肥胖犬多多运动，但在整个 10 周的减肥计划中，每天记录的平均步数却没有增加[9]。作者认为这是由于宠物主人的行为导致的，但这也可能是由于宠物身体情况限制的，或是运动不耐受或者习得性行为。另一项研究也未能证明体重减轻后运动有所增加，这可能表明肥胖症减轻也不会改变宠物主人或者犬的行为习惯[40]。宠物主们经常主观地认为他们的宠物是"活跃的"，日常经常运动，但这很少得到科学研究的证实[10]。目前还没有研究评估如何使超重宠物坚持运动。然而，一项关于人类和宠物共同减肥的运动计划的研究表明，犬和它们的主人可以同时减肥[50]。有趣的是，与宠物一起锻炼的主人的减肥效率与单独锻炼的主人没有什么不同。

7.6　实用建议

- 在不限制能量摄入的情况下增加运动，在临床上不太可能显著出现体重减轻。
- 运动可以在热量限制期间保持瘦体重。
- 体育活动对全身健康的益处可能比减肥的益处更大，但任何活动都应该考虑到疾病合并症或预期的对健康的不利影响，如骨关节炎、心血管健康，或运动耐受性。

- 任何建议都应该强调运动的距离远近，而不是持续的时间，因为对正常（可能是肥胖）犬来说，运动的距离远近是能量消耗的主要决定因素。
- 大多数宠物主人遛犬都没有达到足够的运动距离，从而无法大幅度增加能量的消耗。每天步行 1 h 只会使每天的能量消耗增加一点点。
- 与步行相比，水下跑步机步行会导致在一定运动距离内能量消耗增加，但要实现效果，频率必须等于或大于宠物主可以为宠物提供的常规步行的频率。
- 喂食频率的改变可能会增加猫的活动量，如果主人的日程允许，增加喂食次数没有任何坏处。
- 对宠物主进行有关运动的培训至关重要，提供的建议必须考虑到主人的生活方式和个人观点。

7.7 结论

在不显著限制热量摄入量的情况下，运动锻炼不太可能显著地减轻体重。因此，运动在肥胖症治疗中的作用应该着重在于对身体健康的益处，例如增加耐力、提高康复质量或针对合并症的治疗，或增强心血管健康。使用兽医建议的辅助的运动方式减肥，如水下跑步机疗法，除非按部就班地照做，否则你的花费和付诸的努力得不到与之对应等价的效果。有关于增加猫科动物运动的方法，以及运动对猫肥胖症的治疗意义等方面的描述都很少。需要积累更多的关于肥胖动物在不同活动期间的能量消耗相关数据，并提前设计此类实验的对照实验。

参考文献

1. Courcier EA, Thomson RM, Mellor DJ, Yam PS. An epidemiological study of environmental factors associated with canine obesity. *Journal of Small Animal Practice* 2010;51(7):362–367.
2. Robertson ID. The association of exercise, diet and other factors with owner-perceived obesity in privately owned dogs from metropolitan Perth, WA. *Preventive Veterinary Medicine* 2003;58(1):75–83.
3. De Feo P. Is high-intensity exercise better than moderate-intensity exercise for weight loss? *Nutrition, Metabolism and Cardiovascular Diseases* 2013;23(11):1037–1042.
4. Lee DC, Sui X, Blair SN. Does physical activity ameliorate the health hazards of obesity? *British Journal of Sports Medicine* 2009;43(1):49–51.

5. Swift DL, Johannsen NM, Lavie CJ, Earnest CP, Church TS. The role of exercise and physical activity in weight loss and maintenance. *Progress in Cardiovascular Diseases* 2014;56(4):441–447.
6. Wadden TA, Webb VL, Moran CH, Bailer BA. Lifestyle modification for obesity: New developments in diet, physical activity, and behavior therapy. *Circulation* 2012;125(9):1157–1170.
7. National Research Council Ad Hoc Committee on Dog and Cat Nutrition. *Nutrient Requirements of Dogs and Cats.* Washington DC: National Academies Press; 2006.
8. Mao J, Xia Z, Chen J, Yu J. Prevalence and risk factors for canine obesity surveyed in veterinary practices in Beijing, China. *Preventive Veterinary Medicine* 2013;112(3):438–442.
9. Warren BS, Wakshlag JJ, Maley M, Farrell TJ, Struble AM, Panasevich MR, Wells MT. Use of pedometers to measure the relationship of dog walking to body condition score in obese and non-obese dogs. *British Journal of Nutrition* 2011;106(S1):S85–S89.
10. Roudebush P, Schoenherr WD, Delaney SJ. An evidence-based review of the use of therapeutic foods, owner education, exercise, and drugs for the management of obese and overweight pets. *Journal of the American Veterinary Medical Association* 2008;233(5):717–725.
11. Lund EM, Armstrong PJ, Kirk CA, Klausner JS. Prevalence and risk factors for obesity in adult dogs from private US veterinary practices. *International Journal of Applied Research in Veterinary Medicine* 2006;4(2):177.
12. Degeling C, Burton L, McCormack GR. An investigation of the association between socio-demographic factors, dog-exercise requirements, and the amount of walking dogs receive. *Canadian Journal of Veterinary Research* 2012;76(3):235.
13. Westgarth C, Christian HE, Christley RM. Factors associated with daily walking of dogs. *BMC Veterinary Research* 2015;11:116. doi:10.1186/ s12917-015-0434-5.
14. Kienzle E, Bergler R, Mandernach A. A comparison of the feeding behavior and the human-animal relationship in owners of normal and obese dogs. *Journal of Nutrition* 1998;128(12 Suppl): 2779S–2782S.
15. Rohlf VI, Toukhsati S, Coleman GJ, Bennett PC. Dog obesity: Can dog caregivers' (owners') feeding and exercise intentions and behaviors be predicted from attitudes? *Journal of Applied Animal Welfare Science* 2010;13(3):213–236.
16. Blaak EE, Wim HMS. Substrate oxidation, obesity and exercise training. *Best Practice & Research Clinical Endocrinology & Metabolism* 2002;16(4):667–678.
17. Borsheim E, Bahr R. Effect of exercise intensity, duration and mode on post- exercise oxygen consumption. *Sports Medicine* 2003;33(14):1037–1060.
18. Pedersen BK, Febbraio MA. Muscles, exercise and obesity: Skeletal muscle as a secretory organ. *Nature Reviews Endocrinology* 2012;8(8):457–465.
19. Havel RJ, Naimark A, Borchgrevink CF. Turnover rate and oxidation of free fatty acids of blood plasma in man during exercise: Studies during continuous infusion of palmitate-1–C14. *Journal of Clinical Investigation* 1963;42(7):1054.

20. Issekutz B, Miller HI, Paul P, Rodahl K. Source of fat oxidation in exercising dogs. *American Journal of Physiology* 1964;207(3):583–589.
21. Wakshlag JJ, Struble AM, Levine CB, Bushey JJ, Laflamme DP, Long GM. The effects of weight loss on adipokines and markers of inflammation in dogs. *British Journal of Nutrition* 2011;106(Suppl 1):S11–S14. doi:10.1017/ S0007114511000560.
22. Slentz CA, Duscha BD, Johnson JL, Ketchum K, Aiken LB, Samsa GP, Houmard JA, Bales CW, Kraus WE. Effects of the amount of exercise on body weight, body composition, and measures of central obesity: STRRIDE—A randomized controlled study. *Archives of Internal Medicine* 2004;164(1):31–39.
23. Avenell A, Brown TJ, McGee MA, Campbell MK, Grant AM, Broom J, Jung RT, Smith WCS. What interventions should we add to weight reducing diets in adults with obesity? A systematic review of randomized controlled trials of adding drug therapy, exercise, behaviour therapy or combinations of these interventions. *Journal of Human Nutrition and Dietetics* 2004;17(4):293–316.
24. Leland MH, Edwards TC, Montant J-P. Abnormal cardiovascular responses to exercise during the development of obesity in dogs. *American Journal of Hypertension* 1994;7(4 Pt 1):374–378.
25. Bach JF, Rozanski EA, Bedenice D, Chan DL, Freeman LM, Lofgren JL, Oura TJ, Hoffman AM. Association of expiratory airway dysfunction with marked obesity in healthy adult dogs. *American Journal of Veterinary Research* 2007;68(6):670–675. doi:10.2460/ajvr.68.6.670.
26. Smith GK, Lawler DF, Biery DN, Powers MY, Shofer F, Gregor TP, Karbe GT, McDonald-Lynch MB, Evans RH, Kealy RD. Chronology of hip dysplasia development in a cohort of 48 Labrador retrievers followed for life. *Veterinary Surgery* 2012;41(1):20–33. doi:10.1111/j.1532-950X.2011.00935.x.
27. Newton PM, Mow VC, Gardner TR, Buckwalter JA, Albright JP. The effect of lifelong exercise on canine articular cartilage. *The American Journal of Sports Medicine* 1997;25(3):282–287.
28. Griffin TM, Guilak F. The role of mechanical loading in the onset and progression of osteoarthritis. *Exercise and Sports Sciences Reviews* 2005;33(4):195–200.
29. Bennell KL, Hinman RS. A review of the clinical evidence for exercise in osteoarthritis of the hip and knee. *Journal of Science and Medicine in Sport* 2011;14(1):4–9. doi:10.1016/j.jsams.2010.08.002.
30. Johnston SA, McLaughlin RM, Budsberg SC. Nonsurgical management of osteoarthritis in dogs. *Veterinary Clinics of North America: Small Animal Practice* 2008;38(6):1449–1470.
31. Impellizeri JA, Tetrick MA, Muir P. Effect of weight reduction on clinical signs of lameness in dogs with hip osteoarthritis. *Journal of the American Veterinary Medical Association* 2000;216(7):1089–1091.
32. Linder D, Mueller M. Pet obesity management: Beyond nutrition. *Veterinary Clinics of North America: Small Animal Practice* 2014;44(4):789–806.

33. Brooks D, Churchill J, Fein K, Linder D, Michel KE, Tudor K, Ward E, Witzel A. 2014 AAHA weight management guidelines for dogs and cats. *Journal of the American Animal Hospital Association* 2014;50(1):1–11.
34. Wakshlag JJ, Struble AM, Warren BS, Maley M, Panasevich MR, Cummings KJ, Long GM, Laflamme DE. Evaluation of dietary energy intake and physical activity in dogs undergoing a controlled weight-loss program. *Journal of the American Veterinary Medical Association* 2012;240(4):413–419. doi:10.2460/javma.240.4.413.
35. Chauvet A, Laclair J, Elliott DA, German AJ. Incorporation of exercise, using an underwater treadmill, and active client education into a weight management program for obese dogs. *The Canadian Veterinary Journal* 2011;52(5):491–496.
36. Mlacnik E, Bockstahler BA, Müller M, Tetrick MA, Nap RC, Zentek J. Effects of caloric restriction and a moderate or intense physiotherapy program for treatment of lameness in overweight dogs with osteoarthritis. *Journal of the American Veterinary Medical Association* 2006;229(11):1756–1760. doi:10.2460/javma.229.11.1756.
37. Vitger AD, Stallknecht BM, Nielsen DH, Bjornvad CR. Integration of a physical training program in a weight loss plan for overweight pet dogs. *Journal of American Veterinary Medical Association* 2016;248(2):174–182.
38. Uribe JH, Vitger AD, Ritz C, Fredholm M, Bjornvad CR, Cirera S. Physical training and weight loss in dogs lead to transcriptional changes in genes involved in the glucose transport pathway in muscle and adipose tissues. *The Veterinary Journal* 2016;208:22–27.
39. Thomas DM, Gonzalez MC, Pereira AZ, Redman LM, Heymsfield SB. Time to correctly predict the amount of weight loss with dieting. *Journal of the Academy of Nutrition and Dietetics* 2014;114(6): 857–861.
40. Morrison R, Reilly JJ, Penpraze V, Pendlebury E, Yam PS. A 6–month observational study of changes in objectively measured physical activity during weight loss in dogs. *Journal of Small Animal Practice* 2014;55(11):566–570.
41. Shmalberg J. Part 1: Canine performance nutrition. *Today's Veterinary Practice* 2014;4(6):72–76.
42. Shmalberg J. Part 2: Canine rehabilitative nutrition. *Today's Veterinary Practice* 2014;5(1):87–90.
43. Scott KC, Shmalberg J, Williams JM, Morris PJ, Hill RC. Energy intake of pet dogs compared to energy expenditure at rest, sitting, and standing. *The WALTHAM International Nutritional Sciences Symposium 2013*. Portland, Oregon; 2013.
44. Shmalberg J, Scott KC, Williams JM, Hill RC. Energy expenditure of dogs exercising on an underwater treadmill compared to that on a dry treadmill. *13th Annual AAVN Clinical Nutrition and Research Symposium*. Seattle, Washington; 2013.
45. Hill RC, Scott KC, Williams JM, Morris PJ, Shmalberg J. Energy required for trotting is inversely proportional to leg length in small dogs. *13th Annual AAVN Clinical Nutrition and Research Symposium*, Seattle, Washington; 2013.
46. Clarke DL, Wrigglesworth D, Holmes K, Hackett R, Michel K. Using environmental and feeding enrichment to facilitate feline weight loss. *Journal of Animal Physiology and*

Animal Nutrition 2005;89(11–12):427.
47. Kappen KL, Garner LM, Kerr KR, Swanson KS. Effects of photoperiod on food intake, activity and metabolic rate in adult neutered male cats. *Journal of Animal Physiology and Animal Nutrition* 2014;98(5):958–967.
48. de Godoy MRC, Ochi K, de Oliveira Mateus LF, de Justino ACC, Swanson KS. Feeding frequency, but not dietary water content, affects voluntary physical activity in young lean adult female cats. *Journal of Animal Science* 2015;93(5):2597–2601.
49. Deng P, Iwazaki E, Suchy SA, Pallotto MR, Swanson KS. Effects of feeding frequency and dietary water content on voluntary physical activity in healthy adult cats. *Journal of Animal Science* 2014;92(3):1271–1277.
50. Kushner RF, Blatner DJ, Jewell DE, Rudloff K. The PPET study: People and pets exercising together. *Obesity (Silver Spring)* 2006;14(10):1762–1770. doi:10.1038/oby.2006.203.

8 临床诊疗中建立体重管理程序

8.1 引言

在一般情况下或社区诊疗医院中,建议兽医和助理们为前来就诊的宠物提供综合的减肥计划。在初始检查和制定减肥计划时,在减肥过程和维持体重阶段均随时提供咨询服务。宠物减肥后的日常维持是至关重要的,这有助于它们保持健康的体状且避免体重反弹。因此,宠物主人一定要彻底改变他们与宠物的相处模式,这样才能持续控制宠物对食物的摄入量和活动量[1,2]。通过减肥初期的会诊提高宠物减肥的成功率,以及后期的随访达到长期的减肥效果[3-6]。

8.2 初次会诊

8.2.1 日粮组成、活动情况和家庭生活史

在第一次体重管理预约之前,请宠物主人填写一份有关日粮组成、活动情况和家庭生活史的表格(DAHHF,附录8.1)。理想情况下,宠物主人按要求在线提交表格,或是在会诊时提交纸质的也行。宠物主人在预约前提交表格,可确保兽医团队有时间在预约前准确地评估相关信息。如果没有提前提交表格,DAHHF 也可以现场填写,但是并不主张这样做,正如人们记录的那样,宠物主人无法回忆起他们宠物所食用的食物、零食或是补充品的确切名称以及实际食用量[7,8]。而且现场填写 DAHHF 的时间可能会造成医院日程的延误,除非这已经算在了预约时间内,或是要求宠物主人提前 10~15 min 到达。

利用 DAHHF 的信息计算宠物每餐饮食、零食(商业食品和自制食品)以及用于药物治疗食品的热量,从而在与宠物主预约见面前计算出每日摄入的总热量。在见面时应检查并核实信息,进一步提问以提示或是提醒宠物主人想起他们可能会遗漏的食物。在回顾饮食历史时,让多个家庭成员来述说宠物食用的食物情况会更准确。许多宠物主可能没有意识到补充剂(如鱼油、软嚼)往往也含有热量,这些也需要计入每日的能量摄入中。

除了了解饮食历史外，DAHHF 还会提示宠物主人提供家庭组成（包括全部的家庭成员和宠物）的重要信息，以及宠物所进行的活动信息。对于家庭组成，应该收集每位家庭成员（年龄和性别）的细节信息，谁负责喂食、治疗和陪宠物运动。还有必要知道其他宠物的年龄、性别、品种和是否绝育等详细信息，以及它们与其他宠物的关系，例如是否有血缘关系，是否一起进食或是一起运动，以及它们相处的是否融洽。

为提高工作效率，会安排兽医护士或助理辅助完成这项工作，他们本应该就有记录此类饮食史的工作经验。如果与宠物主见面的时候没完成 DAHHF，让宠物主人在初次会诊后提供更详细的信息也可能会有所帮助。如果宠物主人在会诊后 24 h 内也没有提供 DAHHF，建议拨打电话或是发送电子邮件提醒。

8.2.2 把握初次会诊的信息

一旦背景信息收集完毕，包括营养评估和当前的体重和体况评分，应该向宠物主人说明对宠物的预期理想的体重（见第 5 章）。可使用不同的体况评分系统，分别是 5 分制、7 分制或 9 分制 [9-12]。然而，最近全球营养委员会世界小动物兽医协会（WSAVA）推荐犬和猫普遍采用 9 分制的体况评分系统和"山寨"版本（见第 5 章），这些在网上是可以获得的，可以用来给宠物主人说明他们宠物目前的体况评分情况及预期目标 [13,14]。

在初次会诊时，建议从多角度拍摄宠物站立时的照片（如背部的、侧面的，见图 8.1）。当宠物达到预期目标体重后，与这些照片进行比较（见第 5 章）。这些照片可以用于多种目的，例如在诊所创建一个展示减肥成功的展板、用于宣传营销或是教育案例。若是使用照片进行宣传或是向公众展示，应在初次会诊时与宠物主人签署一份授权协议书。附录 8.2 提供了用于允许使用照片的授权协议书示例。

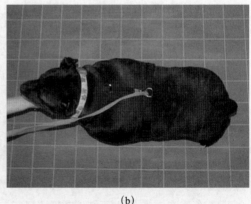

(a)　　　　　　　　　　(b)

图 8.1　初次会诊期间拍摄的照片。当宠物达到目标体重时，可以和这些照片比较
（图片由利物浦大学皇家犬体重管理诊所提供）

许多宠物主人希望在宠物减肥期间为其提供定制食品。这需要准备一份常用的人类食品、商业化食品或是自制食品原料的能量含量清单,这个清单可以当面交给宠物主人,也可以在会诊结束后通过电子邮件发送。宠物主人每天在家使用每日饮食限额表(见附录8.3),可以清楚地说明宠物主人如何使用每日使用饮食限额表,以避免为宠物提供过多的能量。当宠物患有其他疾病时(如限制脂肪或蛋白质的摄入)需要进行营养干预时,食品限量表和个体治疗方案就要加以适当的调整。提供给宠物主人一个每周的食物记录表也是十分必要的(见附录8.4),这里记录着食物供应量和消耗量、运动情况、"招供"(比如偶尔获得的额外食物,可能是宠物自己偷走的或是宠主奖励的)、行为举止和其他可以被记录的情况(如其他疾病)。

在会诊结束时,应向宠物主人提供减肥入门工具包,包括一系列的工具用于帮助他们完成减肥的过程。如果可能,可以提供免费的样品或是具体产品的推荐。建议包括:
- 罐装或是干的食品样品。
- 牵引绳(考虑医院品牌)。
- 精确到克的秤,当宠物主人不方便使用秤称量食物时,可提供适当尺寸的量杯(如1杯、1/3杯、1/4杯、1/8杯)。
- 热量含量明确的食物,以及饲喂量说明。
- 玩具和益智玩具喂食器。
- 可提供的其他相关物品,包括垃圾袋、食物碗等。

通常情况下,生产各类减肥用处方粮的制造商很乐意为此提供他们的产品样品。

除上述提到的这些,还应考虑以组织小组会议的形式提供进一步的帮助,在会议上请宠物主人分享他们的经验。在这些会议期间,兽医工作人员可以通过简短的教育讲座传授简单的体重管理技巧。倘若宠物不存在行为问题,应鼓励犬的主人带着他们的宠物一起参加,而对于猫主人来说,不带着宠物效果更好。这样的会议可能会给其他宠物主人提供友情帮助,他们也在为相同的问题而挣扎烦恼,比如说如何处理宠物的乞食和寻找食物的行为,以及增加游戏活动的建议。如果有可能的话,除了具有减肥计划的宠物主人参加,那些已经减肥成功(或是正在保持体重过程中)的宠物主人们参加会议对他们也是很有帮助的。这使得客户之间能够分享好的实践经验,同时也使宠物主人加深自我认识,在减肥阶段结束后有必要继续保持该计划。

如果无法进行面对面的会议,或是宠物主人对这种形式不感兴趣,可以考虑组织在线论坛(由体重管理小组成员主持)或是群发电子邮件。这样可能达到非常好的效果,但是需要花费时间去管理,这就需要从人员和成本的角度考虑了。最后,这些做法也可以为宠物主人提供相互交流的机会,比如为信念坚定的宠物主人们组建散步小组。这些都很容易建立起来,因为建立和维护这些事情的动力来自宠物主人们,而不是医院诊疗。

小贴士
- 建议宠物主人在家中完成 DAHHF 表格,而不要在办公室完成,以便在首次咨询前获得更准确的饮食历史信息。
- 请指定的兽医护士或助手研究所有食物中的能量含量,计算当前总能量的摄入。
- 向宠物主人提供一份图文并茂的表格,以显示当前的身体状况和预期的最终目标。
- 医院在初次会诊时为肥胖宠物拍摄照片,将来与达到理想体重后的照片进行比对。
- 让宠物主人签订一份照片授权协议书。
- 为宠物主人提供每日饮食限额表和每周食物记录表。
- 在会诊结束后,提供一份减肥工具包,包括推荐的食物样品。
- 考虑建立一个互助小组,为宠物主人提供鼓励并分享减肥成功经验。

8.3 随访和复查预约

最好在与宠物主人首次预约见面 24 h 后,通过电话或电子邮件与宠主联系,以确保他们感到满意或是没有任何直接的顾虑。大多数宠物主人在第一次预约后,宠物主人可能要考虑很多问题,经常会出现混乱或在实施新制度时出现一些问题。联络不仅是因为宠物主人对其重视,而且有助于宠物主人遵守减肥方案,以防他们随后退出[16]。

然后,根据诊所日常管理流程、宠物主人认为方案的可用性和方案进展情况,每隔 2~4 周定期组织一次复诊。理想情况下,应建议宠物主人在最初阶段每 2 周复查一次,因为这段时间需要反复沟通并修改方案。如果最初的食物分配方案是正确的,那么在这个阶段,所有的肥胖宠物的体重都应该减轻一些。这种最初的复查是与宠物主人确认方案是否成功的关键,并可向宠主说明这是值得的。对于正在减肥的宠物,在诊所内复查体重可以与家里的体重检查交替进行,主人在家里称一下宠物体重,然后与诊所联系,定期更新体重信息。以作者的经验,这种方法可能会导致不能长期遵守减肥计划,应该慎重使用,并且只在宠物减肥计划的后期使用。

在临床上,使用相同体重秤定期进行体重检查仍然是确保减肥成功的首选方法。此外,还应为宠主提供免费使用诊所体重秤的服务,以便在办公期间随时检查宠物的体重。体重复查预约应安排在宠物主人离开诊所之前。提前预约有助于长期的随访和使宠物主人遵守减肥方案[16]。理想情况下,所有的预约都应提前24 h确认,体重检查预约的时间可能比其他预约的时间更短。考虑让兽医护士来为预约登记。这类临床护士需要接受相关的培训,以审核宠物每天摄入的总热量,确定当前的体重和体况评分,并计算体重减少的比例。如果体重下降超出了推荐范围(见第5章),可能是宠物主人没有按照计划进行,或者是宠物已经达到了理想的体重和身体状况,建议兽医进行随访。

可以通过体重、体况评分、体型(使用皮尺)和照片等多种方式监控宠物减肥的进展。体重测量本身就可以用来确定减肥的效果,并且应该是被用于决策的主要测量结果[17]。相反的是,体况评分的变化是缓慢的(在9分制的表中,每个体况评分单位通常需要减少5%~15%的体重)[18],因此,体况评分最好定期进行评估而不是每次就诊都进行一次。但是,使用皮尺测定的结果可能不够准确[19],尤其是当测量的人不同时,最好补充其他的测量方法获得的结果用以说明体况的变化。

对于大多数肥胖的犬猫来说,减肥阶段可能持续至少6个月,有些则持续超过1年[20]。在这段时间,如果减肥进度停滞(如在2次检查之间体重没有减少或是增加),绝大多数情况下需要对减肥方案进行调整。通常需要减少食物的供给,一般是少量的(例如,最多分配5%~10%)。如果是干粮并使用精确到克的秤称量,这可以非常精确地调整食物供给量;但使用湿粮或是用杯子进行测量时,很难做出准确的调整[21,22]。在减肥进展过快的情况下(例如,2次检查发现每次体重变化大于每周3%),可以考虑少量增加食物摄入量,但以作者的经验,这种情况是很少见的。减肥进展一般在开始的3~6个月内最快,但不可避免地会在以后变得具有挑战性。如果宠物主人保持着积极性并乐于坚持这个计划,较慢的减肥速度也是可以接受的(比如每周减少0.3%~0.5%)(激励办法见框8.1)。

> **框8.1 特别关注激励措施**
>
> 减肥和维持目标体重的过程对于大多数宠主来说是巨大的挑战,在这个过程中不少人逐渐失去给宠物减肥的信心,因此很多减肥过程停滞不前。因此,制定一个为达到关键节点的客户及其宠物实施奖励和肯定制度,对宠物主人坚定对减肥的信念是有帮助的[1,2]。根据可用资源和个人实践确定确切的奖励分数,建议设置以下4个节点。
>
> (1)体重减轻6%。有证据表明,在减肥计划中,达到这一目标后健康状态得到明显改善[3]。可以是一个简单的奖励,如一枚奖章或是证书。不过,对于动物在达到这一目标时,奖励可以灵活,可以考虑玩具奖励。

(2)达到目标的一半(例如,如果要求总减重为20%,则为开始减重10%)。这是一个很好的节点,比开始时更接近目标,可以为宠物主人提供坚持计划的动力。在这一阶段,宠物主人可能正经历着不忍心忽视宠物乞食行为[4],所以一个益智的喂食器或者类似的刺激可能是一个比较合适的奖励。

(3)完成目标3/4。在这一阶段,体重减少可能比较缓慢且挑战性增加。此时的奖励有助于在最后阶段激励客户。既然能达到这一目标,宠物主人可能会更有动力继续坚持下去,免费提供一些食物可能是对他们迄今为止所有的努力的一种重要的激励。减肥用处方粮的制造商可能愿意为这些宠物提供免费的食物样品。

(4)达到目标体重。可以参考那些已经讨论过的类似情况。

参考文献

1. Webb TL. Why pet owners overfeed: A self-regulation perspective. In: *Companion Animal Nutrition Summit: The Future of Weight Management*. Barcelona, Spain: Nestle Purina; 2015, pp. 89–94.
2. Murphy M. Obesity treatment: Environment and behavior modification. *The Veterinary Clinics of North America: Small Animal Practice* 2016;46:883–898.
3. Marshall WG, Hazewinkel HAW, Mullen D et al. The effect of weight loss on lameness in obese dogs with osteoarthritis. *Veterinary Research Communications* 2010;34:241–253.
4. Bissot T, Servet E, Vidal S et al. Novel dietary strategies can improve the outcome of weight loss programmes in obese client-owned cats. *Journal of Feline Medicine and Surgery* 2010;12:104–112.

小贴士

- 首次会诊后24 h内联系宠物主人,解决任何紧急问题。
- 为宠物主人提供免费的体重秤,以便在上班时间他们可以随时为宠物称重。
- 建议每2~4周复查一次体重,如果进展良好,可以间隔更长的时间再复查。
- 最好在宠物主人离开诊所之前确定复查的时间,并且提前24 h确认。
- 兽医护士需要经过适当的培训后,才能从事体重复查工作。
- 整个减肥过程中需要调整食物的摄入量,推荐使用精确度到克的秤,以提高食物供给量的准确性和精确性。

8.4 维持期的复诊

一旦减肥的宠物达到理想体重和状态,应该提供资料用于确认其成功。可以将宠物减肥后的照片与之前的照片进行比较。考虑为宠主提供一个证书,详细记录初始体重、体况评分和体脂百分比(通过形态测量法进行估计和量化),并与最终的数值做比较。也可以考虑为宠物提供一些品牌商品来表明他们在医院里成功减肥了,比如头巾、项圈或是牵引绳。在医院里展示减肥宣传板是向其他宠物主人展示该减肥计划成功并鼓励他们加入的绝佳方法。每月或是每一季度突出展示成功减肥的宠物(例如,医院的"明星减肥者"),确保挑选出减肥最成功的宠物和它的主人。灵活设置评选获奖者的标准,使大家以各种原因获奖。这种方式确实有必要,这个奖励可以及时地给予特定的客户,以帮助他们在需要的时候得到鼓励。

在完成减肥阶段时,还应准备一个信息包提供给宠物主人,以便为宠物体重维持阶段做准备。这个信息包应该包括更新的喂食建议,以帮助宠物成功地维持新取得的健康身体状况。宠物主人(实际也包括兽医专家)通常会认为减肥过程已经完成,可能会让最初增加体重的坏习惯再次出现。许多研究表明,成功达到目标体重的犬和猫,大约有一半在这时体重反弹[5,6]。因此,应该花时间在体重管理过程实施跟踪计划。最初,应该恢复每2~4周开展一次体重维护期检查,在这一时期食物摄入量是逐渐增加的,每次的增加量很小(每次最高不超过5%~10%)。一旦目标体重维持至少2个连续的复诊后,检查的时间间隔可以逐渐延长(例如,每4~8周,然后每3个月,之后至少每6个月)。可以考虑给这类体重复查项目提供部分优惠价格,尤其是由兽医护士承担的此类工作。但是,如果遇到体重增加或是不符合标准的情况,建议与兽医一起制定新的减肥计划。

有许多原因会导致特定的动物在成功达到目标体重后又反弹,但一个关键因素是维持体重期间食物的类型。事实上,在随访时发现,保持为动物提供与减肥期间相同的食物(例如,特制的体重管理食物),尽管增加了供给量用以维持体重,体重反弹的可能性也要低20倍[5]。鉴于这类食物的额外成本,宠物主人可能不愿意继续长期饲喂这类食物。如果可行的话,可以以折扣价提供此类食品给宠物主人,以确保他们认为继续使用这类食品是物有所值的。

小贴士
- 为宠物主人提供证书、宠物减肥前后的照片或印有医院名称的礼物,以证明宠物减肥成功。
- 考虑在诊所大厅摆放一块减肥成功案例宣传板。
- 创建一个宠物体重维持阶段的信息包,为宠物主人在宠物体重维持阶段做准备。
- 在体重维持阶段也要如减肥阶段一样时刻注意体重,以确保尽早发现体重反弹并纠正。
- 建议开始需频繁地复查体重,随后间隔可以逐渐延长,持续至最多每6个月复查一次，以保持稳定的体重。
- 建议在宠物体重维持阶段继续为宠物主人提供定制的用于体重管理的食物，并以折扣价提供给他们。

8.5 宣传营销

在实践中实施体重管理服务时，建议准备简洁的营销材料，如分发材料、小册子或在医院网站信息页面宣传。信息应该包括项目的背景、过程（例如，初次会诊后进行的一系列体重复查）、激励措施（比如，减肥入门包）。所有的员工都要接受适当的培训，以回答关于减肥计划的各种咨询问题，诊所还可以举办活动，邀请当前参与减肥项目的客户参加活动，并向新客户做宣传。这些活动可能包括定期的团体散步、诊所发起的社区散步或者是趣味跑步，并且可以作为那些决心要开始减肥旅程的宠物主人们的动员支持会议。对于养猫的人来说，可以建立一个猫活动区（猫活动区示例见框8.2）。

框8.2　受到特殊关注的微恙猫咪健身房（由 Kenneth J. Lambrecht 创建）

Kenneth J. Lambrecht

起源

微恙猫咪健身房的开发灵感源于已发布的一项公共数据，数据显示理想体重的犬比肥胖的犬寿命增加近2年[1]。当搬去一个新的地点时，中西部（Midwestern）兽医机构的老板将这些信息应用到新空间的开发和设计中，重点在于控制理想的体重和营养调控。专门给猫的那部分也被设计成符合美国猫从业者协会（American Association of Cats Practitioners）的黄金级猫友训练的要求。

创造设备，优化空间

诊所从接待区（图8.2）上方的连接2条猫道的隧道开始，然后在大楼的第2层增加了1个完整的健身房和猫的寄宿区（图8.3）。"微恙猫咪攀爬山"成为第一个专为猫设计的空间。它是由2块夹板制成的，夹板上留有孔洞，方便猫从内部通道到下面的隐藏空间。它的角度允许垂直爬升和逃生（图8.4）。接着，增加了一堵攀岩墙和一个快速的猫练习轮[2]。随后，该机构创建了"微恙猫咪私人健身房"，供害羞的寄宿的猫通过隧道进入私人公寓去运动（图8.5）。增加了更多的爬梯和高架猫道，允许猫在整个房间范围内活动（图8.6）。

图 8.2 接待猫通道。医院接待区上方有 2 条与隧道相连的猫道
(经美国威斯康星州麦迪逊市西城兽医中心许可使用该图片)

图 8.3 微恙猫咪健身房。第 2 层建设为健身房和寄宿区
(经美国威斯康星州麦迪逊市西城兽医中心许可使用该图片)

图 8.4 微恙猫咪攀爬山。"微恙猫咪攀爬山"由 2 块夹板组成,夹板上有为猫从内部通道到下面的隐藏空间而开的孔。它是倾斜的,允许垂直攀登和逃生。背景中还可以看到攀岩墙
（该图片经美国威斯康星州麦迪逊市西城兽医中心许可使用）

图 8.5 布格私人健身房
（该图片经美国威斯康星州麦迪逊市西城兽医中心许可使用）

图 8.6 增加了更多的爬梯和高架猫道,使猫能够在整个房间的活动范围内活动
(该图片经美国威斯康星州麦迪逊市西城兽医中心许可使用)

理想的体重管理计划"拯救宠物减肥大赛"和"猫之夜"

微恙猫咪健身房是该诊所举办的一年一度8周龄拯救宠物减肥大赛（PRFR）的一个重要部分。比赛期间,当地的宠物主人和拯救人员加入超重猫参与的体重管理计划。这些猫每减少1 lb体重,诊所就会向当地救援机构提供捐赠,有些赠品是销售公司作为赞助商捐赠的产品。参与奖励包括在初始称重期间提供的入门奖品,以及有机会以折扣价购买活动监视器、体重秤和自动喂食器,这部分所得的款项会捐赠给参与的救援机构。参与拯救宠物减肥大赛的猫如果是新手或者比较害羞,可以单独预约安排或是通过每月的"猫咪之夜"活动来健身房活动。

这些活动包括关于适当喂养和运动的教育讨论,具体到猫的行为方面。那些成功参与减肥计划的猫主人可以帮助管理这些活动,并为那些刚开始为猫减肥的人提供支持。微恙猫咪健身房可以让猫和人在减肥的过程中觉得更加有趣。这家诊所还发现,通常情况下,经过不断的健身房锻炼和参加每月1次的"猫咪之夜"活动,这些猫咪变得更加活泼并喜欢社交。

> **参考文献**
>
> 1. Kealy RD, Lawler DF, Ballam JM et al. Effects of diet restriction on life span and age-related changes in dogs. *Journal of the American Veterinary Medical Association* 2002;220(9):1315–1320.
> 2. One Fast Cat. Cat Exercise Wheel. Available from: http://onefastcat.com. Accessed March 1, 2016.

在康复/运动医疗机构的专业人员,例如,美国兽医运动医学与康复学院认证医师(DACVSMR),认证犬康复医师(CCRP),或认证犬康复理疗师(CCRT)他们的共同努力下,可帮助减肥计划进一步取得成功(见第7章)。如果不是隶属于同一医院,不同机构之间可以建立合作转诊制度。如果是在一个单一的医院网络内提供康复或是锻炼服务,康复/运动医疗服务机构对运动锻炼的评估可以作为初始减肥评估的一部分。另外,减肥评估也可以推荐给那些对于主要咨询康复/运动医学服务的超重或肥胖宠物进行。与当地的宠物敏捷性训练中心或水上运动训练机构联系是另一个为宠物主人提供资源的机会,这可以鼓励宠物主人通过运动与他们宠物互动而不仅仅是通过喂食。其中一些组织可能有兴趣联合赞助一个针对刚起步的肥胖或超重宠物的活动。

社区中也可能有机会与人体减肥诊所、营养师或是体育馆合作,这些机构在为人们提供减肥计划方面具有专业知识。考虑为这些机构提供兽医支持,重点是那些自身也超重的宠物主人。

小贴士

- 在诊所和网上为宠物主人提供营销材料。
- 培训所有员工正确回答关于减肥计划的问题。
- 考虑为现有客户举办活动来促进减肥或维持健康体重,同时也向新客户做宣传。
- 与专业人员一起确定当地动物康复/运动医学计划,在自己的医院内建立协作转诊或合作服务。
- 联系当地组织,如敏捷性训练中心或水上运动训练机构,为宠物主人提供额外的运动资源。
- 为当地的人体减肥机构提供兽医服务,重点是自身也超重的宠物主人。

8.6 经济因素的影响

经济拮据可能会影响宠物主人为其宠物参加减肥计划。减肥疗程很可能持续

6~12个月甚至更长的时间,而每2~4周支付全额诊疗费[20],这对许多宠物主人来说是难以承受的。那些为宠物购买了医疗保险的人情况可能会好些,但是有些保险可能不包括与减肥项目有关的福利。因此,诊所应考虑通过利用兽医护士预约,创建打折的体重复诊费用或收取肥胖咨询套餐费用来免费提供部分或全部的体重检查项目。可以根据实践情况来确定此类服务包含的内容,选项包括以下任何一项或全部。

- 单独与兽医和/或护士进行体重检查。
- 减肥食品(或折扣券)。
- 精确度为克的秤(如果提供)。
- 奖品和激励措施。
- 常规化验(例如定期血细胞化验、电解质检测及尿液化验)。
- 获取网上资源。
- 后续的体重维护项目。

在实践中可以考虑提供不同成本价位的套餐,从而使客户能够选择适合他们的选项。与减肥套餐不同,体重管理"套餐"可能是另一种选择(例如,预付一定数量的体重复诊费用)。在这些复诊完成之后,可以单独支付额外的预约,购买更多的套餐,或者客户可以转换到正式的体重管理套餐。那些完成了为期2年基于减肥计划的生活方式的人群中有95%表示愿意支付持续的减肥干预费用,包括每月联系教练[23]。

从财务和营销的角度来看,诊所可能会考虑把这个套餐作为医院为招揽顾客而亏本的服务[24]。尽管这可能违反常理,但因为体重管理服务造成少量的收入损失可能会鼓励宠物主人坚持执行减肥计划,然后定期去医院就诊。这不仅有助于使客户与医疗机构保持联系,而且更频繁地去医院就诊可能会通过增加销售额(例如更新处方、安排保健预约、购买预防药物)使医疗机构受益。总之,这种减肥检查可以及早发现其他疾病,然后需要在诊所进行随访。

与引入诊所的任何新计划或服务一样,应在设置客户最终消费之前进行成本效益分析。这一过程涉及制定一份与实施成本相对比的诊所收益列表。收益可能是有形的(例如,诊所收入)或无形的(例如,当宠物穿着品牌头巾/皮带/项圈时带来的客户口碑、媒体关注、医院广告)。诊所的较大费用之一可能是减肥入门工具包或奖励。降低该工具包成本的方法是批量购买物品、获得物品捐赠以及从宠物食品制造商处索要食品样品或优惠券。

如果宠物主人安排这些就诊的目的是讨论宠物新出现的与体重不相关的症状,那么实施包括折扣体重检查预约的系统可能会给常规设置带来挑战。与宠物主人进行日程安排和参加体重复诊谈话时与客户沟通是最关键的。医院工作人员应始终将预约称为体重检查预约,并向客户重申其并不包括全面的身体检查。可能与这些宠物主人互动的医院工作人员包括以下人员。

- 负责预约的客户服务代表。
- 在核查预约登记时接待客户的客户服务代表。
- 兽医护士或助手,为主治兽医保定宠物或完成体重复诊。
- 为宠物进行评估的主治兽医(如果兽医护士未能完成评估)。
- 在就诊后将宠物送出医院时的客户服务代表。

鼓励这些工作人员将这些来访称为减肥期或减肥维护期的访问,以减少客户对其预期范围的任何困惑。

小贴士

- 考虑收取打折的减肥复诊费和/或创建减肥项目和复诊捆绑的套餐。
- 考虑在医院设立为招揽顾客而折本出售商品的项目。
- 应该使用成本效益分析来确定项目收益与实施成本之间的关系。
- 所有与这些预约有关的工作人员必须明确告知宠物主人是体重复查或维护期的预约,并且不包括一般健康或不良宠物全面体检。

8.7 结论

每个建立体重管理计划的兽医诊所都应首先创建资源,以帮助其服务宠物主人的宠物实现减肥并最终尽可能有效地保持理想体重。初步信息调查表(DAHHF)、饮食限额表,每周食物记录表以及明确说明的理想体重和体况评分目标将帮助兽医工作人员和宠物主人交流有关喂养建议的信息。在减肥之前和之后的宠物照,减肥入门套件和减肥完成证书可以激励宠物主人坚持遵守计划。还可以通过定期体重复诊和开放诊所体重秤的方式确定方案的合理性,以便于宠物主人检查宠物在体重评估期间的进展情况。明确的喂养说明有助于维持理想体重,并且应长期保持至少每 6 个月做一次体重检查。与减肥有关的营销材料和医院活动可以作为向社区宣传的广告,也可以与其他兽医或人类医学专业人士合作。

最后,应通过成本效益分析来评估建立体重管理计划所涉及的经济成本,以确保该计划从客户和诊所的角度都切实可行。

附录 8.1　饮食、活动和家庭生活史记录表（DAHHF）

以下示例表格是基于美国利物浦大学皇家犬重量管理诊所和田纳西大学兽医学院营养服务所使用的表格。

新病例宠物信息调查表

该问卷的主要目的是在您前往诊所之前为我们提供有关您的宠物的初步信息。我们将使用该调查表作为咨询的基础。它还将使我们能够根据您的宠物的情况量身定制合适的体重管理方案。

请咨询所有家庭成员（如果可能），并尽可能如实地回答问题。如果不确定任何某具体问题的含义，请不要担心。我们可以在您访问期间进行解答。另外，如果您有任何问题不想回答，请随时将其留白。

第一部分：关于您的宠物

宠物名字：_____　　姓：_____

种类：□犬　□猫
性别/状态：□正常雌性　□绝育雌性　□正常雄性　□绝育雄性

品种：_____　出生日期：_____　年龄：_____（年/月）

您的宠物是几岁绝育的？
_____

您的宠物是从哪里来的（如饲养员、救援人员、收容所）？
_____

当时您宠物的年龄？年龄：_____（年/月/周）

请让我们知道您的宠物在过去 12 个月中患有的任何疾病：
_____
_____
_____
_____
_____
_____

请让我们知道您宠物之前任何的其他医疗问题：
_____
_____
_____
_____

您的宠物有保险吗？□是　□否
　　如果有,监护人是谁？_____
您是否注意到宠物排尿有任何变化？□是　□否
　　如果是,请描述_____
目前您的宠物食欲好吗？□是　□否
　　如果否,请描述_____
您的宠物的食欲最近有变化吗？□是　□否
　　如果是,请描述_____
您的宠物是否经历了不正常的体重增加或体重减轻？□是　□否
　　如果是,请描述_____

您的宠物当前的体重是多少？_____　□磅　□千克　称量日期：_____
您的宠物：_____　□低于理想体重　□理想体重　□高于理想体重
如果您的宠物体重过低或过高,您宠物的理想体重是多少？
_____　□磅　□千克

当前的跳蚤 / 壁虱 / 心丝虫病预防 (名称和实施频率)
举例:品牌 X 控制跳蚤 (10~22 lb 犬):每 4 周嚼 1 次 (最后一次是 2018 年 1 月 10 日)

1. _____
2. _____
3. _____

当前服用的药物 / 补充剂 (每天的名称和剂量)
例如:泼尼松 (5 mg 片剂):每日 2 次 1½ 片剂

1. _____
2. _____
3. _____
4. _____
5. _____
6. _____
7. _____
8. _____

第二部分:关于您的家庭
每个家庭成员的详细信息

	年龄	性别	关系	职业
您				
其他成员				

如有需要,请填写以下内容

请描述主要的宠物主人未提供的任何护理(如日托、遛犬、寄宿、宠物看护)

_____
_____
_____
_____

请提供其他宠物的详细资料

名字	种类	品种	年龄	性别	是否绝育 (是/否)	是否存在亲缘关系 (是/否)

如有需要,请填写以下内容

剩下的问题要为将要前往诊所的宠物填写

第三部分:喂养主食和零食

谁喂养这只宠物? _____
您的宠物通常在哪里喂养(例如,厨房、洗衣房、犬窝、室外)? _____
您的宠物在平常的一天有几餐(放入碗中的食物)?
□1次　□2次　□3次　□4次　□被忽视　□其他: _____
您的宠物和家中其他宠物是共用一个碗喂食吗? □是　□否
　　如果是,请描述: _____
宠物之间有食物竞争吗? □是　□否
　　如果是,请描述: _____
您的宠物除了自己的食物以外还能获得其他食物吗? □是　□否
　　如果是,请描述: _____
您的宠物有没有翻找过垃圾桶? □是　□否
　　如果是,您的宠物多久翻找一次垃圾桶? _____

完整的饮食记录(市售饮食)

食物类型	品牌	风味	每餐进食量	每天进食次数	饲喂起止日期	停止饲喂原因

如有需要,请填写以下内容

完整的饮食记录(自制饮食)

饮食/成分类型	制作方法	每餐进食量	每天进食次数	饲喂起止日期	停止饲喂原因

如有需要,请填写以下内容

完整的零食记录(市售零食、咀嚼物等)

零食类型	品牌	风味	规格	每天食用量	饲喂起止日期	停止饲喂原因

如有需要,请填写以下内容

完整的零食记录(人类食品、餐桌剩饭剩菜等)

零食类型	成分	分量	每天饲喂次数	饲喂起止日期	停止饲喂原因

如有需要,请填写以下内容

您是否使用食物/零食来管理药物或补品?□是　□否
　　如果是,请描述:＿＿＿＿＿＿＿＿＿＿＿＿＿＿＿＿＿＿＿＿

您的宠物喝什么?
　　□自来水　□井水　□瓶装水　□牛奶　□茶　□其他 ＿＿＿＿＿

您如何测量宠物食物的量?
　　□估计量　□量杯/勺　□称重　□其他 ＿＿＿＿＿＿＿＿＿

您的宠物乞求食物吗?□是　□否
　　如果您的宠物乞求食物,它多久这样做一次?
　　　　□全天　□一天几次　□每天1次或2次　□每天少于1次

第四部分:犬生活环境

从 1~10 的范围,您的犬有多活跃?＿＿＿＿＿＿
  1= 非常不活跃。整天都在睡觉。运动时很少嬉戏。
  10= 非常活跃。完全地享受运动,经常在室内或室外活动。

您多久锻炼一次您的犬?
  ☐ 每天不止 1 次 ☐ 每天 1 次 ☐ 几天 1 次
  ☐ 从不 ☐ 其他:＿＿＿＿＿＿＿＿＿＿＿＿＿＿＿＿

您在哪里锻炼您的犬?
  ☐ 一层的住宅 / 公寓 ☐ 一层以上的住宅(楼梯)
  ☐ 后院 ☐ 花园 ☐ 当地人行道
  ☐ 当地公园 / 游乐区 ☐ 郊野公园 ☐ 田野或林地
  ☐ 遛犬公园 ☐ 其他:＿＿＿＿＿＿＿＿＿＿

平均而言,您的犬每次锻炼多长时间?
  ☐ 10 min 或更少 ☐ 10~20 min ☐ 20~60 min
  ☐ 1 h 或更多 ☐ 其他:＿＿＿＿＿＿＿＿

在运动过程中,您的犬会
  ☐ 与散步者一起玩 ☐ 与球 / 棍棒 / 其他玩具一起玩
  ☐ 不拴绳 ☐ 用可伸缩的牵引绳 ☐ 其他:＿＿＿＿＿＿

您和您的犬住在哪里?
  ☐ 带院子 / 花园的公寓 ☐ 不带院子 / 花园的公寓
  ☐ 有院子 / 花园的房子 ☐ 无院子 / 花园的房子
  ☐ 带院子 / 花园的简易别墅
  ☐ 不带院子 / 花园的简易别墅
  ☐ 其他:＿＿＿＿＿＿＿＿＿＿＿＿＿＿＿＿＿＿＿＿

您的犬大部分时间都花在哪里?
  ☐ 室内 ☐ 室外 ☐ 室内外

您的犬可以去户外吗?
  ☐ 从不 ☐ 有栅栏的院子 / 花园 ☐ 无栅栏的院子 / 花园
  ☐ 仅步行

您的犬什么时间能在户外度过?
  ☐ 工作日 ☐ 周末

您的犬每天要独处几个小时?
  ☐ 0~2 h ☐ 2~5 h ☐ 5~8 h ☐ 8 h 以上

从 1~10 的等级范围,您的犬受过良好训练的等级?＿＿＿＿＿＿＿
  1 = 未经训练(无基本服从)
  10 = 训练有素(例如,敏捷性训练、表演训练、技巧)

第五部分　猫生活环境

从 1~10 的范围,您的猫有多活跃？ _____
　　1 = 非常不活跃。很少练习,做到最低限度。
　　10 = 非常活跃。经常锻炼,总是在旅途中,游戏时精力非常充沛。
您的猫大部分时间都在哪里度过？
　　□室内　　　　　　　　□室外　　　　　　　　□室内外
您的猫咪在户外活动受限吗？ □是　□否
　　如果是,您什么时候将猫关在房子里？
　　　　□夜间　　　　　　□宠物主上班时
　　　　□房主在家中的一天　□其他:_____
您的猫每天在户外度过多少时间？
　　□少于 1 h　　　□1~3 h　　　□3~6 h　　　□6 h 以上
您的猫会狩猎吗？ □是　□否
您的猫会带回来活的猎物吗？ □是　□否
您的猫每天要独处几个小时？
　　□0~2 h　　　□2~5 h　　　□5~8 h　　　□8 h 以上
您的猫会不会藏在屋子里？ □是　□否
　　如果是,您的猫藏在哪里？_____
您的猫会自己玩吗？　□是　□否
　　如果是,是多久？　□每天 1 次　　□每天 1 次以上
　　　　　　　　　　□每周 1 次　　□每周 1 次以上　　□每周 1 次以下
您会和您的猫玩吗？　□是　□否
　　　　　　　　　　□每天 1 次　　□每天 1 次以上
　　　　　　　　　　□每周 1 次　　□每周 1 次以上　　□每周 1 次以下
您的猫喜欢什么玩具？
　　□球　　　　　□钓鱼玩具　　　□卷起的报纸　　　□发条玩具
　　□猫薄荷　　　□其他:_____
您的猫每天睡眠几个小时？
　　□少于 12 h　　□超过 12 h
您和您的猫住在哪里？
　　□带院子/花园的公寓
　　□不带院子/花园的公寓
　　□有院子/花园的房子
　　□无院子/花园的房子
　　□带院子/花园的简易别墅
　　□不带院子/花园的简易别墅
　　□其他:_____
如果您家中有 1 只以上的猫,这些猫共享以下哪些资源？
　　□喂食站　　　□水站　　　□休息处　　　□垃圾箱

附录8.2　客户照片发布许可表

如果将宠物的照片用于市场推广或向公众展示,宠物主人应在初次预约时签署本发布表格(此表格经美国新泽西州廷顿瀑布市红岸兽医医院批准使用)。

本人_____特此允许____(医院名称)____在任何已知的或以下涉及的媒体中永久使用我的宠物的医疗故事、我自己的照片和我的宠物的照片,并且我已知情他们有权自行决定更改上述照片。我已知情他们可以选择此时不使用这些照片或我的宠物的病历,但以后可以自行决定使用。____(医院名称)____也保留在不另行通知的情况下中断使用这些照片的权利。我已了解,如果在____(医院名称)____医院网站发表或发布了我自己或我的宠物的照片,则任何计算机用户都可以下载该照片。因此,我同意对因使用上述照片____(医院名称)____的医院任何人员所引起的任何索赔免责。

签字:_____

日期_____

(医院名称)代表:_____

附录8.3　饮食限额表

猫犬每日饮食限额表是根据每个病例的情况量身定做的。诊所应该推荐目前饮食和一些低热量的饮食,以帮助宠主清楚地看到他们如何使用他们的饮食限额表,并避免在饮食中提供过多的热量。

(这些例子是经美国田纳西大学兽医学院营养服务许可修改的。)

每日饮食限额 —— 猫

宠物名字:_____

每天最大能量摄入:_____

为了保持饮食均衡,食物的热量不应超过每日总热量摄入的10%。只要您不超过建议的每日最大能量,就可以混合搭配以下零食。下面列出的饮食方案并不受特定制造商的支持或结果的保证。

当前饮食

列表基于您当前正在饲喂的食物。如果您希望喂食其他零食,请考虑单个产品的能量含量,以确保您的宠物不超过建议的每日饮食限额。

品牌/产品名称	每种食物的能量含量

低热量商业食品

以下食物的能量含量通常允许在每日限额内每天可喂食多片。

品牌/产品名称	每种食物的能量含量/(cal/个)
品牌香脆猫零食(任何口味)	1.1
X品牌健齿零食(任何口味)	1.25
X品牌健齿猫零食	1.3
X品牌猫零食(任何口味)	1.5
X品牌猫零食(任何口味)	2
X品牌含鸡肉和土豆粉猫零食	3

零食分配玩具

下面列出的玩具可能对提高猫的日常活动量很有用。可以在这些玩具中放置全部食物或部分日常食物,以使它们对猫产生健康的环境刺激。

品牌/产品名称
X品牌隧道猫喂食器
X品牌猫玩具
X品牌猫咪甜筒
X品牌互动猫玩具

最后更新日期:_____

每日饮食限额——犬

名字：_____

每天最大能量摄入：_____

为了保持饮食均衡，食物的热量不应超过每日总热量摄入的10%。只要您不超过建议的每日最大能量，就可以混合搭配以下零食。下面列出的饮食方案并不受特定制造商的支持或结果的保证。

当前饮食

列表基于您当前正在饲喂的食物。如果您希望喂食其他零食，请考虑单个产品的能量含量，以确保您的宠物不超过建议的每日饮食限额。

品牌/产品名称	食物尺寸	每种食物的能量含量

低热量商业食品

以下食物的能量含量通常允许在每日限额内每天可喂食多片。

品牌/产品名称	食物尺寸	每种食物的能量含量/(cal/个)
X品牌膨化食品	特小	4
X品牌奶酪味食品	迷你	4.5
X品牌香脆食品(任何口味)	小	6
X品牌耐嚼食品	小	9
X品牌软(烤食品)	1号	12
X品牌饼干	中	20

商业健齿食品（兽医口腔卫生委员会批准）

当按照产品标签上的描述给予时，下面列出的食品可以减缓牙菌斑和牙垢的形成。*请注意：您宠物的饮食限额可能低于许多牙科食品选择的能量含量。

品牌/产品名称	食物尺寸	每种食物的能量含量/(cal/个)
X品牌健齿食品	特小	26
X品牌口腔护理健齿食品	小	63
X品牌薄荷味呼吸健齿食品	大	99

最后更新日期：_____

附录 8.4　每周饮食日记

这使宠物主人可以记录所提供的食物、所消耗的食物、运动量和"检讨"（例如，食用了其他食物的情况，无论是动物偷走的还是主人提供的食物），以及行为和其他任何观察结果（例如其他疾病）。这些例子是经英国内斯特的利物浦大学皇家犬重量管理诊所许可而修改的。

每周饮食记录表

宠物名字：_____

日期	食物供给量	食入食物量 （全部/一半/无）	运动 运动量/运动频率	总体意见	陈述	问题/意见

饮食周数_____
请指出何时开始使用新的袋子、托盘、罐头、食品袋或盒。

参考文献

1. Murphy M. Obesity treatment: Environment and behavior modification. *Veterinary Clinics of North America: Small Animal Practice* 2016;46:883–898.
2. Linder D, Mueller M. Pet obesity management: Beyond nutrition. *Veterinary Clinics of North America: Small Animal Practice* 2014;44: 789–806, vii.
3. German AJ, Holden SL, Gernon LJ et al. Do feeding practices of obese dogs, before weight loss, affect the success of weight management? *British Journal of Nutrition* 2011;106 (Suppl 1):S97–S100.
4. German AJ, Holden SL, Wiseman-Orr ML et al. Quality of life is reduced in obese dogs but improves after successful weight loss. *Veterinary Journal* 2012;192:428–434.
5. German AJ, Holden SL, Morris PJ et al. Long-term follow-up after weight management in obese dogs: The role of diet in preventing regain. *Veterinary Journal* 2012;192:65–70.
6. Deagle G, Holden SL, Biourge V et al. Long-term follow-up after weight management in obese cats. *Journal of Nutritional Science* 2014;3:e25.

7. Schoeller DA. Limitations in the assessment of dietary energy intake by self-report. *Metabolism* 1995;44:18–22.
8. Archer E, Pavela G, Lavie CJ. The inadmissibility of what we eat in America and NHANES dietary data in nutrition and obesity research and the scientific formulation of national dietary guidelines. *Mayo Clinic Proceedings* 2015;90:911–926.
9. Laflamme D. Development and validation of a body condition score system for cats: A clinical tool. *The Feline Practice* 1997;25:13–17.
10. Laflamme DP. Development and validation of a body condition score system for dogs. *Canine Practice* 1997;1997:10–15.
11. Thatcher CD, Hand MS, Remillard RL. Small animal clinical nutrition: An iterative process. In: *Small Animal Clinical Nutrition*, 5th ed., MS Hand, CD Thatcher, RL Remillard et al. (eds.). Topeka, KS: Mark Morris Institute; 2010, pp. 3–21.
12. German AJ, Holden SL, Moxham GL et al. A simple, reliable tool for owners to assess the body condition of their dog or cat. *Journal of Nutrition* 2006;136:2031S–2033S.
13. World Small Animal Veterinary Association Global Nutrition Committee. Body condition score chart dogs. 2013. Available at: http://www.wsava.org/sites/default/files/Body%20 condition%20score%20chart%20dogs.pdf. Accessed February 28, 2016.
14. World Small Animal Veterinary Association Global Nutrition Committee. Body condition score chart cats. 2013. Available from: http://www.wsava.org/sites/default/files/ Body%20condition%20score%20chart%20cats.pdf. Accessed February 28, 2016.
15. German A, Heath S. Feline obesity: A medical disease with behavioral consequences. In: *Feline Behavioral Health and Welfare*, 1st ed., I Rodan, S Heath (eds.). St. Louis, MO: Saunders; 2015, pp. 148–161.
16. American Animal Hospital Association. Compliance: Taking quality care to the next level: A report of the 2009 AAHA compliance follow-up study; 2009:18. Available from: https://secure.aahanet.org/eweb/images/student/pdf/Compliance.pdf.
17. National Heart, Lung, and Blood Institute with The National Institute of Diabetes and Digestive and Kidney Diseases. *Clinical Guidelines on the Identification, Evaluation, and Treatment of Overweight and Obesity in Adults: The Evidence Report*. Publication No. 98-4083. Bethesda, MD: National Institutes of Health, National Heart, Lung, and Blood, Institute; 1998.
18. German AJ, Holden SL, Bissot T et al. Use of starting condition score to estimate changes in body weight and composition during weight loss in obese dogs. *Research in Veterinary Science* 2009;87:249–254.
19. Baker SG, Roush JK, Unis MD et al. Comparison of four commercial devices to measure limb circumference in dogs. *Veterinary and Comparative Orthopaedics and Traumatology* 2010;23:406–410.
20. German AJ, Titcomb JM, Holden SL et al. Cohort study of the success of controlled weight loss programs for obese dogs. *Journal of Veterinary Internal Medicine* 2015;29:1547–1555.

21. German AJ, Holden SL, Mason SL et al. Imprecision when using measuring cups to weigh out extruded dry kibbled food. *Journal of Animal Physiology and Animal Nutrition* 2011;95:368–373.
22. Murphy M, Lusby AL, Bartges JW et al. Size of food bowl and scoop affects amount of food owners feed their dogs. *Journal of Animal Physiology and Animal Nutrition* 2012;96:237–241.
23. Jerome GJ, Alavi R, Daumit GL et al. Willingness to pay for continued delivery of a lifestyle-based weight loss program: The Hopkins POWER trial. *Obesity* 2015;23:282–285.
24. Hess J, Gerstner E. Loss leader pricing and rain check policy. *Marketing Science* 1987;6:358–374.

9 病例分析

9.1 病例1：一只腊肠犬椎间盘突出手术后成功减肥并维持体重——风险因素评估的重要性

MORAN TAL、CLAUDIA WONG 和 ADRONIE VERBRUGGHE

9.1.1 既往病史

一只6岁的雄性绝育腊肠犬因背痛入院寻求治疗。核磁共振成像显示颈椎间盘在C5和C6之间压迫脊髓，之后进行了腹侧切口椎体切除术。由于该机构的神经科服务中心将患犬评估为肥胖体况，因此建议它减肥以减轻关节压力，患犬的主人也同意通过该机构的临床营养服务机构来减肥。

9.1.2 病情评估

在营养咨询前（该时间点称为 T_0），患犬恢复良好。根据世界小动物兽医协会营养评价指南[1]进行的营养筛查评估显示，其唯一不健康的就是肥胖问题。宠物主人说，曾经它每天吃2次宠物店品牌的犬干粮①（宠物食品标签信息：代谢能 ME 478 kcal/杯或3 900 kcal/kg，数值根据分析保证值计算得出；蛋白质≥108.2 g/1 000 kcal；脂肪≥51.3 g/1 000 kcal；粗纤维≤14.1 g/1 000 kcal）。由于它的食物是用一个回收利用的容器来测量的，所以无法计算准确的每日能量摄入量，2个装满的容器被宠物主估计为略低于1个标准杯（8 oz,250 mL）。该犬每天进食2个容器那么大量的犬粮。此外，为了激发它对食物的兴趣，还喂它吃了煮熟的鸡肉（约112 kcal/4 tbsp 或每天60 g），以及没有标注能量含量的商业零食。全天还提供烟熏牛棒骨（每根骨头重约96 g,能量约198 kcal）和生蔬菜。总的来说，它每日能量摄入量中37.5%以上（>300 kcal/d）来自零食和不均衡的餐桌食物。

它的体况评分（BCS）[2]被评估为9/9（根据临床医生的经验估计体脂大约为50%），但肌肉状况评分（muscle condition score，MCS）[3]由于脂肪覆盖范围过大而无法评估。它的初始体重（BW）为15.8 kg，而它的评估理想体重为

---

① 渴望，原味犬粮（干粮），产地为加拿大埃德蒙顿。

9.88 kg［(15.8 kg×50% 瘦体重)/80% 瘦体重］[2,4]。

由于患犬的体况评分(BCS)大于 5/9，因此，需要进一步的营养评估[1]。腊肠犬这个品种、公犬绝育、中年都是导致其肥胖的自身因素。饮食因素包括高能量含量食物、过量食用零食以及人类的食物，加之不当的食物能量评估。后来，患犬的活动能力也很低，尤其是处于恢复期活动能力就更低。

9.1.3 治疗方案

从一些选择中筛选出一种用于减肥的兽医处方粮①、干性犬粮和罐头犬粮配方(干性犬粮和罐头犬粮的营养成分分别为：ME 360 kcal/100 g 和 56 kcal/100 g；蛋白质 94.4 g/1 000 kcal 和 121.8 g/1 000 kcal；脂肪 27.8 g/1 000 kcal 和 35.8 g/1 000 kcal；粗纤维 7.5 g/1 000 kcal 和 17.9 g/1 000 kcal；L-肉碱 83.3 g/1 000 kcal 和 62.7 g/1 000 kcal)。根据理想体重 9.88 kg(390 kcal/d)计算每日能量摄入，使用 1.0 的减肥维持能量需求(maintenance energy requirement, MER)系数(MER=$70 \times 9.88^{0.75} \times 1.0$=390 kcal/d)[4]。

每天给该犬喂食 3 次，以增加饱腹感，减少讨食行为。鼓励它的主人用喂食玩具(漏食球②)喂他，以丰富和增加活动度。对于零食，推荐低热量、兽医治疗用犬饼干③(ME 15 kcal/片)和生蔬菜，不超过每日能量摄入的 10%，以保持营养均衡[5]。所有其他零食和人类食物都被严格禁止。要求宠物主用厨房克秤称量每一餐的重量，以确保准确的喂食量。为确保每周能缓慢持续的减重 1%~2%[6]，2% 是建议的最大安全减肥量[4]，建议每 2 周称重 1 次，在家庭兽医的诊治中始终使用相同的体重秤。根据文献，如果体重变化低于 0.5%，建议每日能量摄入的减少幅度为 5%~20%[4,7]。因此，食物量供给应根据体重的减轻情况进行调整。

9.1.4 短期和长期随访

T_0 后的 12 d，患犬体重下降了 0.9 kg(BW 14 kg)(平均每周 3.5%)，如果这种速度持续稳定下去，则考虑调整食物摄入量来逐渐减重。10 d 后，其体重没有变化。调整饮食计划，将每日能量摄入减少到 360 kcal/d(减少 8%)，分为每天 4 餐而不是 3 餐。经过这一调整，接下来的 4 周多体重一直保持持续的下降状态，然后趋于平稳。

T_0 后的 2 个多月(68 d)，患犬和宠物主来预约复查(图 9.1)。虽然最近的体重趋于进入平台期，但犬的身体状况已从 BCS9/9 改善到 8/9(BW 12.2 kg)。宠物主表示其活动能力和能量水平有所改善，机构建议宠物主继续采取相同的治疗方案。然而，在复查后的 13 d，患犬体重却增加了 0.2 kg，并没有更低水平。随着天气转暖，宠物主人相信可以增加该犬的活动量，因此仍然是沿袭原有的治疗方案。

---

① 皇家犬能量控制处方粮(干粮和罐头)，产地加拿大安大略省。
② Kong 公司，产地科罗拉多州戈尔登市。
③ 普瑞纳冠能处方粮，低热量零食，产地加拿大安大略省。

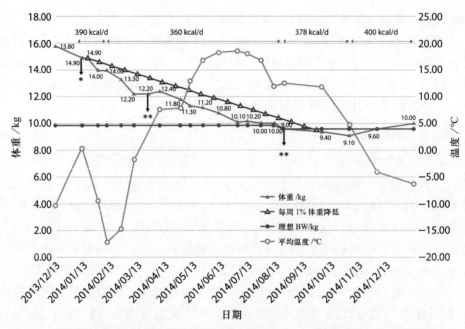

图 9.1 体重下降趋势与理想体重、平均每周体重下降 1% 的初始体重和平均温度相对应。
随着时间的推移,热量摄入量显示在图的顶部
* 2014 年 1 月 17 日:减肥计划开始(T_0)。
** 2014 年 3 月 26 日、2014 年 8 月 22 日(Tm):预约复查。

正如预测的那样,随着运动量的增加(如每天散步的次数和持续时间的增加),该犬在整个春季都在持续减重状态。体重减轻的速度各不相同。然而,在 T_0 后大约 6 个月,该犬体重达到 10.0 kg。

在下一次预约复查时(T_0 后 7 个月),该犬体重又减少了 0.4 kg(图 9.1),9.6 kg 体重被评估为理想的 BCS 和 MCS。采用相同的 2 种食物,计算出一个长期的体重维持计划(时间点称为 Tm),每日能量摄入增加 5%,达到 378 kcal/d[4]。机构建议每月更新体重。

1 个半月后,该犬在 2 次体重记录中减掉了 0.5 kg,因此它的体重维持计划也相应地修改为能量摄入量 400 kcal/d,增加了 6%。

Tm 后 4 个月,患犬体重增加了 0.5 kg,由于体重持续增加,在冬季剩余时间内,考虑到活动量较低,它的每日能量摄入量调整回 378 kcal/d。在减肥计划开始 1 年后,该犬以这种能量摄入保持其体重水平不变。

9.1.5 讨论

如第 1 章所述,肥胖的危险因素包括动物自身因素和人为因素,如饮食因素、身体活动、宠物主人态度和居住环境特征。在本病例中,营养评价显示存在许多这

些易感因素。该犬是绝育的雄性腊肠犬（易感品种）[8,9]，最初是以非处方高脂肪食物喂养的，没有合适的测量容器（在计算每日分量时不精确且不准确）[10]。此外，当犬出现讨食行为或被认为对食物不感兴趣时，宠物主会奖励它营养不均衡的食物。由于背痛，该犬也几乎不活动，这又更降低了它的能量消耗。可见消除以上一些易感因素对减肥至关重要。

在制订减肥计划时，能量限制，即减少每日能量摄入量，对于实现负能量平衡最为重要[4,11]。为了达到这一目的，应选择一种为减肥而配制的处方粮，因为维持性食物的能量限制可能缺乏一些必需的营养素，如蛋白质、必需氨基酸、必需脂肪酸、矿物质和维生素。

减肥的关键营养因素包括但不限于降低能量密度，从而减少脂肪[4,12]。增加总膳食纤维（total dietary fiber, TDF）含量[4,13]，也会稀释能量密度，并通过减缓胃肠道转运时间来增加饱腹感[14]。增加蛋白质（当限制为80%的静息能量需求时，应满足>60 g/1 000 kcal）[4,15]，对于减肥期间的瘦体重维持[15]及其显著的热效应[16]非常重要。如前所述，考虑到减肥的关键营养因素，选择了用于减肥的处方粮。此外，根据美国饲料管理协会[17]的建议，这种宠物食品也适合长期的体重管理，因为其被认为是全面且均衡的成年动物食物。

使用食物玩具喂食和锻炼有助于宠主感到有积极参与减肥过程。此外，特别是在天气适宜的情况下（图9.1），促进日常锻炼不仅能促进体重减轻，保持肌肉量，而且还能加强犬与人之间的关系[18]。减重/增重率的变化也证明了定期更新体重、与客户沟通以及根据患宠的个体需求调整减肥计划的重要性。

> **学习要点**
>
> （1）营养评估对于识别和消除肥胖症的危险因素非常重要。
> （2）根据患宠和宠物主的需要量身定制个性化减肥计划。
> （3）建议经常重新评估减肥计划，并根据患宠、客户和环境因素多次频繁监测并个性化调整。
> （4）与客户沟通是确保长期保持稳定体重和生活质量的关键要素。

参考文献

1. Freeman L, Becvarova I, Cave N et al. WSAVA nutritional assessment guidelines. *Journal of Small Animal Practice* 2011;52(7):385–396.
2. Laflamme DP. Development and validation of a body condition score system for dogs. *Canine Practice* 1997;22(4):10–15.

3. Baldwin K, Bartges J, Buffington T et al. AAHA nutritional assessment guidelines for dogs and cats. *Journal of the American Animal Hospital Association* 2010;46(4): 285–296.
4. Brooks D, Churchill J, Fein K et al. 2014 AAHA weight management guidelines for dogs and cats. *Journal of the American Animal Hospital Association* 2014;50(1):1–11.
5. Yaissle JE, Holloway C, Buffington CA. Evaluation of owner education as a component of obesity treatment programs for dogs. *Journal of the American Veterinary Medicine Association* 2004;224(12):1932–1935.
6. Laflamme DP, Kuhlman G. The effect of weight loss regimen on subsequent weight maintenance in dogs. *Nutrition Research* 1995;15(7):1019–1028.
7. Fragua V, Barroeta AC, Manzanilla EG, Codony R, Villaverde C. Evaluation of the use of esterified fatty acid oils enriched in medium-chain fatty acids in weight loss diets for dogs. *Journal of Animal Physiology and Animal Nutrition* 2015;99:48–59.
8. Lund EM, Armstrong PJ, Kirk CA, Klausner JS. Prevalence and risk factors for obesity in adult dogs from private US veterinary practices. *International Journal of Applied Research in Veterinary Medicine* 2006;4(2):177.
9. Edney AT, Smith PM. Study of obesity in dogs visiting veterinary practices in the United Kingdom. *Veterinary Record* 1986;118(14):391–396.
10. German AJ, Holden SL, Mason SL et al. Imprecision when using measuring cups to weigh out extruded dry kibbled food. *Journal of Animal Physiology and Animal Nutrition* 2011;95(3):368–373.
11. German AJ. The growing problem of obesity in dogs and cats. *Journal of Nutrition* 2006;136(7 Suppl):1940S–1946S.
12. Borne AT, Wolfsheimer KJ, Truett AA et al. Differential metabolic effects of energy restriction in dogs using diets varying in fat and fiber content. *Obesity Research* 1996;4(4):337–345.
13. Jackson JR, Laflamme DP, Owens SF. Effects of dietary fiber content on satiety in dogs. *Veterinary Clinical Nutrition* 1997;4:130–134.
14. Jewell D, Toll P. Effects of fiber on food intake in dogs. *Veterinary Clinical Nutrition* 1996;3:115–118.
15. Hannah SS, Laflamme DP. Increased dietary protein spares lean body mass during weight loss in dogs (ABSTRACT). *Journal of Veterinary Internal Medicine* 1998;12:224.
16. Hoenig M, Waldron M, Ferguson D. Effect of a high-and low- carbohydrate diet on respiratory exchange ratio and heat production in lean and obese cats before and after weight loss. *Compendium on Continuing Education for the Practising Veterinarian: North American Edition* 2006;28(4):71.
17. Association of American Feed Control Officials (AAFCO). *2014 Official Publication AAFCO*; 2014.
18. Westgarth C, Christian HE, Christley RM. Factors associated with daily walking of dogs. *BMC Veterinary Research* 2015;11:116.

9.2 病例 2：一只成年绝育雌性史宾格猎犬的病态肥胖症
MEGAN SHEPHERD

9.2.1 既往病史

一只 8 岁的雌性绝育史宾格犬，以下称"猎犬"，被送到兽医教学医院（VTH）骨科，针对右后肢跛脚症状进行评估与诊疗。当时这只猎犬足有 50.9 kg，合 112 lb，体况评分为 9/9。骨科检查期间，由于该猎犬过度肥胖和因检查产生的应激引起了呼吸困难。导致骨科检查中止，并将其送至氧气笼中恢复。

兽医教学医院骨科服务中心要求通过营养咨询协助猎犬减肥。与宠物主人讨论了患犬身体康复状况，并得到主治骨科专家的建议。然而，由于宠物主人工作繁忙，时间安排不上，以及距离康复器材较远等原因，还没有进行正式的身体康复。

营养师约见了宠物主人以了解清楚该宠物的过往详细饮食情况。第一位宠物主人早上喂食 1 杯老年犬专用的干粮①（74 kcal/ 杯，已经饲喂 1~2 年时间），并称该猎犬只有在把医生建议吃的别的食物吃完后才吃这些犬粮。此外，还给它喂食了一种洁牙零食②（小中型犬专用款 42 kcal/ 根），1 片意大利辣香肠（脱脂95%，每片能量约 10 kcal），半条低脂 / 部分脱脂马苏里拉奶酪棒（每条能量为约 70 kcal）。午餐时，第一位宠物主人喂它 4~5 片市售③鸡肉或者火鸡肉。晚餐时，第二位宠物主人喂给其同样的老年犬专用干粮，半块低脂 / 部分脱脂马苏里拉奶酪（在那之前，第二位宠物喂了一整根马苏里拉奶酪）。在犬服用药物、关在犬笼里和两餐之间向主人乞食时，主人还会喂它多种零食。零食包括煮熟的鸡蛋（每天 2 个全蛋 ± 1 个鸡蛋白）、肉罐头（该犬最近停吃）、烤鸡胸肉、意大利面、通心粉和奶酪、马苏里拉奶酪棒（半根）、意大利辣香肠或更多老年犬用干粮。宠物主人把可利用的剩菜剩饭喂给该犬。该猎犬当前这种饮食已经有 1~2 年的时间了。营养师估算出当前的饮食每天至少提供了 1 300 kcal 能量。宠物主近期也尝试着喂食减肥用处方粮④1~2 周，以促进减肥。刚开始该猎犬还吃这些处方粮，然而很快就失去了兴趣。

据宠物主人讲，他们并没有让该犬对此款减肥处方粮进行换粮过渡。

---

① 唯一鸡肉米饭配方老年犬粮（374 kcal/ 杯，没有标每千克中含的能量信息）。
② 宝路洁齿棒清新口气零食（42 kcal/ 个，没有标每千克中含的能量信息）。
③ 奥斯卡美公司。
④ 希尔斯 r/d 犬处方粮（干粮：饲喂水平 242 kcal/ 杯，2 968 kcal/kg；罐头：饲喂水平 257 kcal/12.3 oz 罐，733 kcal/kg）。

9.2.2 病情评估

目标体况评分设定为5/9。基于它目前体重是理想体重的140%,营养师评估该猎犬理想的体重是 36.4 kg(约为 80 lb),(50.9 kg/140)× 100 = 36.4 kg[1]。这个评估很可能并不准确,因为可能受到 9 分制的体况评估体系的限制,以及这只猎犬的体重高出理想值的 40%。英国史宾格猎犬品种的体重标准一般小于 50 lb。营养师告知主人们,一旦达到理想的体况评分(BCS)(5/9),该猎犬的真实理想体重将得到更准确的评估。

营养学家根据静息能量需要计算公式(RER,$70 \times 36.4 \text{ kg}^{0.75}$)[1],估计该猎犬的维持能量需要(MER)为每天 1 000 kcal。目标设定为该犬每个月减掉体重 2~4 kg(意味着每周减掉初始体重 50.9 kg 的 1%~2% = 0.5~1 kg)营养师建议在减肥期间每月进行体重核算,至少直到出现健康的减肥趋势。营养师建议宠物主人考虑每次称体重使用相同的秤。考虑到该犬巨大的体格,他们当地兽医服务中心的地磅可能更加实用。

9.2.3 减肥计划

营养师设定了最初的饮食目标使这只猎犬适应健康的饮食结构。由于进食了太久的人类食物,营养师期望这只猎犬能够在一定时间内接受适应健康的饮食结构。营养师与宠物主人们讨论如何让该犬能在 4 周时间里尽快适应这"理想的"减肥餐。营养师提出了好几个减肥用处方粮的选择,也劝说宠物主人重新考虑最初的那个针对减肥的治疗方案,假如那个方案还可以被重新慢慢引入进来。

宠物主人选择重新使用之前喂食的那款减肥用处方粮。营养师建议慢慢地过渡到这款减肥用处方粮中。营养师建议早餐和晚餐时取消 1/8~1/4 杯干老年犬粮,并以 1/8~1/4 杯减肥用处方粮(干粮或罐头)代替。刚开始的时候可以临时使用一些适口性好的配料有助于换粮,比如金枪鱼汤、鸡汤罐头,1~2 茶勺碎牛油,1 汤勺的奶酪(融化以更好地粘在食物上),或者 1 汤勺的帕尔玛奶酪。宠物主人可以把少量的减肥处方罐头加在午餐肉上,然后渐渐地增加减肥用处方罐头的数量,减少午餐肉的量。一旦宠物适应了减肥用处方粮,主人就按照特定的指导方案继续。

对于减肥用处方粮,主人可以选择 3.75 杯干粮,或是 3.5 个罐头,再或者是 1.5 杯干粮外加 2 个罐头。宠物主在这 3 种搭配中选择一种,每天的食物总量分成 3 等份(早餐、午餐和晚餐),宠物主人也可以把食物分成 2 等份(早餐和晚餐),然后午餐食用推荐的零食。

建议使用特定的零食方案(表 9.1)。零食限制在不超过每天摄入热量的 10% 或者每天不超过 100 kcal。告知宠物主人每种零食的推荐用量最多满足每天能量需要的 10%。因此,如果每天饲喂多于 1 种零食,宠物主人就得把每种的量分别降低。举个例子,如果宠物主人想要每天喂 4 种零食(比如,喂药或关入犬笼时用于引诱,或是两餐之间),宠物主人每天就要饲喂每种推荐量的 1/4。

表 9.1 该猎犬的零食选择

零食名称	每天的量(每种大约 100 kcal)
花椰菜或者彩色卷心菜	2 杯
切成薄片的生西葫芦	5 杯
胡萝卜丝	2 杯
煮沸,沥干,无盐或生的青豆	2 杯
煮沸,沥干,无盐或生的小胡萝卜	20 个
煮沸,沥干,无盐或生的花椰菜	4 杯
不含黄油/盐的爆米花	3 杯
最好是低钠的脱脂马苏里拉奶酪	1 根
低钠的 95% 无脂香肠、牛肉或猪肉	10 片(最多 100 kcal)
低钠午餐肉(鸡肉或火鸡)	3~4 片(评估每片的热量,最大 100 kcal)
无酱的煮熟通心粉	1/2 杯

注:食品的营养成分使用美国农业部营养数据库确定的标准。

营养师建议宠物主人鼓励犬自发活动,建议可将食品玩具作为鼓励自发运动的工具,提供了几种食物玩具选项[①],并简要介绍了它们与干性减肥用处方粮的配合使用。营养师鼓励宠物主人今后重新考虑对犬进行身体康复,并分享了接受身体康复的肥胖犬的 YouTube 视频。

如果该猎犬的健康状况发生变化,和/或如果该猎犬每月减掉的体重少于 4 lb 或超过 8 lb,则应指示宠物主人与他们的私人兽医联系。如果该猎犬的体重超出了建议的范围,那么营养师需重新检查饮食是否合适,然后根据需要调整每日总热量。

9.2.4 定期复查

在第 4 周电话回访的时候,宠物主人说该猎犬在最初的这 4 周里完全适应了这个治疗方案的饮食结构,而且他们也看到该犬的身体形态发生了变化。宠物主人对目前的结果非常满意。在第 4 周的电话回访时,该犬在兽医诊疗中心称重是 96 lb。第一个主人准备午餐时,该猎犬正在吃减肥用处方粮(3¾ 杯,908 kcal/d)和

---

① KONG 食物玩具、Buster 食物立方体、Aikiou 活动食物中心、Kyjen 犬拼图、Ethical Pet 找食玩具拼图、StarMark Bob-A-Lot 互动宠物玩具。

一小块未测量的烤鸡肉。犬的主人们称，他们已经削减了所有其他零食，但无法提供具体数量。他们也表示，这只犬运动更好了，更有腰型了。主人对这一进展感到满意。在第5个月的电话回访中，主人描述这只犬仍然进食减肥用处方粮，并且增加了运动量。主人没有及时更新体重数据，但是说明友们都说该犬比以前瘦了。犬主人们描述说该猎犬最近饮水量增加，背部的毛发脱落一些，而且已经与兽医预约了时间对它近期身体的变化进行评估。

总之，主人们对于猎犬的进步感到由衷的高兴。该猎犬的减肥是一个开放型病例。在开始4个月减掉的16 lb体重符合每月4~8 lb的目标范围。营养师为宠物主人能显著地改变该犬的体况而感到自豪和高兴。

这个病例表明，彻底熟悉了解饮食历史以及管理方法的重要性。而且在达到减肥目的同时最大限度减小宠物主人的压力和工作量。

9.2.5 讨论

这只猎犬由于能量摄入及消耗之间极度不平衡，尤其是过度的能量摄入导致了它过度肥胖。减肥在多方面使该犬受益。它开始的时候在兽医服务中心表现出后肢跛脚，解决过度肥胖将会缓解肌肉骨骼系统本不应该承受的压力。宠物主人表示该犬在最初的4个月内，在达到理想的体重之前，就已经改善了走路的状态。该猎犬的主人成功实施了这项减肥计划，其中包括对当前饮食和喂养管理（即主人的生活方式）的关注。

仔细考虑喂养管理和日常的饮食日程表有助于制定最大限度地减轻宠物主人压力的减肥计划。根据作者的经验，不当的换粮方式可能会使宠物对新食物的接受产生负面影响，尤其是从吃人类食物转变到吃新食物中。作为一个开放型病例，瘦身的全过程轨迹并不能完全清楚，根据作者的经验，很多过度肥胖的宠物总会有一个瘦身的平台期。当减肥达到平衡状态时，应评估饮食依从性，并调整推荐量。此外，作者使用的方法是评估理想体重的一种方式，该方法影响着减肥过程中的目标能量需求。另一个评估理想体重的方式是从当前的体况基础上减掉超重的40%。归根结底，减肥是从初始计划开始，初始计划的成功将由持续的监控和根据需要重新制定饮食计划来决定。

学习要点

(1) 不要因被肥胖的犬击败而失去信心，因为它的体重很大需要减肥。

(2) 仔细考虑现在和过去的饮食，喂养管理有助于微调减肥建议。

(3) 你可能需要花点时间把一只病态肥胖的犬过渡到适当的饮食。主要以人类食物为食物的犬，不太可能很快就开始减肥。

(4) 提供详细建议。除了特定的饮食建议外,还包括特定的零食建议。

(5) 复查、复查、复查,并提醒宠物主人做好工作。

参考文献

1. Brooks D, Churchill J, Fein K et al. 2014 AAHA weight management guidelines for dogs and cats. *Journal of the American Animal Hospital Association* 2014;50(1):1–11.

9.3 病例3:肥胖小动物的麻醉问题
LYDIA LOVE

9.3.1 既往病史

一只做过绝育手术的7岁家养短毛猫,由于最开始的巨结肠病及随后的慢性便秘而接受了全身麻醉结肠次全切除术。患宠已经使用处方粮[①]、乳果糖和复方西沙必利保守治疗了6个月。然而,便秘还是经常复发且频繁地住院治疗,宠物主不得不考虑选择手术的方式干预治疗。

9.3.2 患猫评估

患猫体重21.8 lb(9.9 kg),体况评分为>9/9(图9.2)。体格检查发现中度牙齿疾病,听诊心脏和肺部声音正常。由于肥胖,很难触诊腹部,但在降结肠中感觉到牢固的大便。

9.3.3 治疗方案

计划全身麻醉进行结肠次全切除术。麻醉问题分为以患宠和手术为导向的问题,对于该患宠而言,主要包括通气不足和低氧血症、亚临床心脏功能障碍、镇静剂和麻醉剂的正确剂量以及疼痛的管理。

肥胖患宠的呼吸系统比正常体重的受试者更不顺应,这是由于脂肪量增加,呼吸方式受限以及肺血容量增加引起的,吸气和呼气流量限制可能会发生,并可能伴有低通气。肥胖患宠以低肺容量呼吸,与正常体重患者相比,功能残存容量(FRC)降低。脂肪质量增加和潮气量减少导致气道关闭,随后肺不张和气体交换不平衡。由全身麻醉药和背卧位引起的剂量依赖性呼吸抑制会加剧通气不足,引起呼吸性酸中毒,并可能导致围手术期的饱和度降低。该肥胖患宠在诱导

---
① 皇家高纤维易消化处方粮,美国皇家宠物食品有限公司。

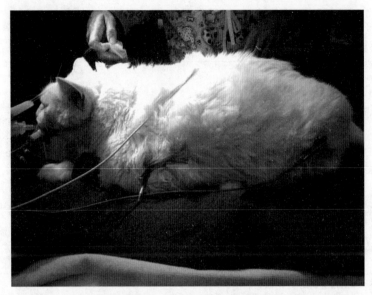

图9.2 一只做过绝育手术的肥胖的7岁家养短毛猫,正在被麻醉

前计划进行预吸氧,以使FRC脱氮,并增加肺中的氧气储存,延长去饱和时间[1]。全麻期间,通过监测呼气末二氧化碳浓度来评估通气。控制通气(手动或机械)用于维持正常的二氧化碳浓度,并密切跟踪脉搏血氧测定。密切客观的氧合监测持续到恢复期。

虽然在犬猫上的描述不如在人类上的描述,但肥胖患宠可能表现出亚临床舒张性心功能不全,需要在麻醉期间密切监测心血管功能,并注意围麻醉期液体负荷。避免使用可引起快速心律失常的药物(尤其是游离麻醉药和抗胆碱能药物),因为当存在舒张功能障碍时,心率加快可能会限制心脏充盈。静脉输注晶体是以手术率计算,但以理想体重为基础。除术中失血过多外,围手术期总液体负荷控制在10~15 mL/kg。

由于肥胖相关的药代动力学参数的变化,镇静药和麻醉药的剂量减少需要计算出来。随着总体重的增加,由于增加的脂肪体重,瘦体重也增加,但不是平行的。大多数麻醉剂和镇静剂都是中度到高度脂溶性的,这会促进它们穿过血脑屏障到达作用部位。因此,肥胖患宠的总体重(TBW)剂量可能导致相对过量。

然而,基于理想体重(IBW)的计算可能会低估患宠的剂量,并且在肥胖的犬猫中还没有建立通用的剂量标准。该患宠的药物计算基于调整后的体重(ABW)[IBW + 0.4(TBW − IBW)]来解释增加的瘦体重[2]。但是,在监测心肺副作用时,将所有麻醉药滴定至有效。结肠次全切除术是一种侵入性腹部手术,需要一定时间的背卧(通常少于2 h)。除了上述肺不张和通气不足的增加外,肥胖患宠的背部卧位可能加重关节疾病,并可能导致与压迫有关的神经和肌肉并发症。手术台上放了额外的垫子,猫的四肢放在一个舒适的位置。最后,预计腹部手术后即刻会出现中度到重度疼痛,围手术期采用多种预防性镇痛方案。

9.3.4 短期和长期随访

患宠超重 40%,初始体重评估为 13.5 lb(6.1 kg)[3]。估计理想体重的原因是,建议的计算 IBW① 的方法尚未在严重肥胖患者中得到验证(BCS>9/9)[3]。药物剂量是根据 ABW(16.8 lb,7.6 kg)计算的。

患宠静脉注射芬太尼(5 mcg/kg),透射脉搏血氧仪探头安装在剃光毛的尾部。附有示波法无创动脉血压袖带和 ECG 导联。使用紧身口罩给患宠预吸氧 5 min,然后用氯胺酮(2 mg/kg)和丙泊酚(1.8 mg/kg)的组合将其麻醉。与氯胺酮共导入可以减少异丙酚的需求,并在诱导时产生更好的血流动力学特征[4]。用部分静脉内技术维持麻醉,包括 100% 氧气中的七氟醚滴定,以芬太尼(10 mcg/kg/h)和氯胺酮(0.6 mg/kg/h)输注进行滴定。考虑硬膜外麻醉,但解剖学标志难以触及。

在手术室,患宠被放置在背卧位,并开始机械通气,采用时间循环、容积控制、限压麻醉的呼吸机。将一个 5 cm H_2O 正呼气末压(PEEP)阀置于呼吸回路的呼气支中。监测包括连续脉搏血氧仪、无创示波血压计、二氧化碳图、血压计、体温计和二导联心电图。通过减少吸入剂和静脉注射麻黄碱(0.03 mg/kg)治疗轻度低血压。患宠的体温始终保持在 100 °F 以上,没有主动外部加热。

手术结束前,中断吸入麻醉,芬太尼和氯胺酮注射液减少 50%。患宠转为侧卧,自主呼吸 100% 氧气。在自发通气期间,潮气末二氧化碳浓度高于 55 mmHg。为了避免低通气导致低氧血症,在恢复期放置一根鼻腔氧气管继续补充氧气(图 9.3)。以 50 mL/kg/min 的速度补充氧气直到第二天早晨。脉搏血氧饱和度透射反射探头保持在尾端,以继续监测氧饱和度。镇痛药物输液在恢复后的前 12 h 之后逐渐减少,通过服用长效丁丙诺啡(0.27 mg/kg/m^2)② 在第二天早上实现了止痛药覆盖范围的减少。长效丁丙诺啡在出院前 24 h、麻醉恢复后 37 h 重复使用。

9.3.5 讨论

体重指数增加与麻醉时间延长、手术时间延长、呼吸和肾脏并发症风险增加、感染风险增加、住院时间延长以及围手术期费用增加有关[5,6]。这些问题很可能与肥胖、小动物麻醉剂患宠有关,预计资源的使用会增加。

---

① 理想体重 = [当前体重 × (100 - 体脂肪百分比)]/0.8。
② 丁丙诺啡,硕腾公司。

图 9.3　肥胖患宠在麻醉恢复室用鼻线补充吸入氧,降低因残余低通气和肺不张而导致低氧血症的风险

通过预测可能出现的并发症,制定规避这些并发症的策略,监测预期和意外并发症的发展,并做好围麻醉期发生问题的准备,这样麻醉的风险可以得以降低。肥胖患宠主要关注的是呼吸功能和麻醉剂和镇静剂的剂量减少。

所有患宠在麻醉诱导前都建议进行预吸氧,以避免去饱和。由于肥胖患宠的代谢需求增加,再加上 FRC 小得多,诱导时会迅速发生去饱和。

该患宠接受预吸氧 5 min,然后用手快速插管并通气。在维持阶段,将患宠放在机械呼吸机上。肥胖人类患者的麻醉后通气策略尚无共识,在犬猫中也没有有关该主题的公开信息。潮气量可能最好基于 IBW 以避免气压或容积伤。加上 PEEP 的肺泡募集动作可减少麻醉期间肥胖人类患者的肺不张并改善顺应性和气体交换[7]。该病例报告的患宠在麻醉期间使用了 PEEP 瓣膜,但未进行肺泡募集操作,这是因为未发现低氧血症发作。停止机械通气后,尽管仍需补充 100% 的氧气,但仍记录到换气不足。为了避免由于持续通气不足和由于残余肺不张引起的通气-灌注不匹配而导致的去饱和,放置了一个鼻氧管,并在整个恢复过程中以连续脉搏血氧仪对患宠进行了客观监测。

最后,根据 ABW 计算该患宠的初始药物剂量。即使对于人类,也没有足够的信息提供有关肥胖症中药物剂量的可靠建议,并且几乎没有对任何种类的肥胖者进行药代动力学-药效学研究。理想情况下,将存在针对特定物种和药物的准则,但如果缺少这些准则,则使用理想体重加上一定百分比的超重可能会导致随着

TBW 的增加而出现的瘦体重的非线性增加。一般而言,最好在监测心肺副作用的同时仔细地滴定麻醉剂以产生作用。

> **学习要点**
>
> 　　对于肥胖和正常体重的受试者,麻醉的安全性不是由某种药物或药物组合决定的,而是由良好的监测和支持治疗所决定的。
>
> 　　(1)通气不足和低氧血症是肥胖猫犬的最常见的麻醉后并发症,应密切监测呼吸系统。
>
> 　　(2)缺乏具体的药代动力学-药效学信息,应根据调整后的体重计算出麻醉药和镇静药的剂量,然后滴定以达到效果。

参考文献

1. McNally EM, Robertson SA, Pablo LS. Comparison of time to desaturation between preoxygenated and nonpreoxygenated dogs following sedation with acepromazine maleate and morphine and induction of anesthesia with propofol. *American Journal of Veterinary Research* 2009 November;70(11):1333–1338.
2. Servin F, Farinotti R, Haberer JP et al. Propofol infusion for maintenance of anesthesia in morbidly obese patients receiving nitrous oxide a clinical and pharmacokinetic study. *Anesthesiology* 1993 April;78(4):657–665.
3. Brooks B, Churchill J, Fein K et al. 2014 AAHA weight management guidelines for dogs and cats. *Journal of the American Animal Hospital Association* 2014 January–February;50(1):1–11.
4. Martinez-Taboada F, Leece EA. Comparison of propofol with ketofol, a propofol-ketamine admixture, for induction of anaesthesia in healthy dogs. *Veterinary Anaesthesia and Analgesia* 2014 November;41(6): 575–582.
5. Higgins DM, Mallory GW, Planchard R et al. Understanding the impact of obesity on short-term outcomes and in-hospital costs after instrumented spinal fusion. *Neurosurgery* 2016 January;78(1):127–132.
6. Ranum D, Ma H, Shapiro FE et al. Analysis of patient injury based on anesthesiology closed claims data from a major malpractice insurer. *Journal of Healthcare Risk Managemant* 2014;34(2):31–42.
7. Aldenkortt M, Lysakowski C, Elia N et al. Ventilation strategies in obese patients undergoing surgery: A quantitative systematic review and meta-analysis. *British Journal of Anaesthsia* 2012 October;109(4):493–502.

9.4 病例4：减肥计划中肥胖的短毛猫肥胖程度估算
CHARLOTTE REINHARD BJØRNVAD

9.4.1 既往病史

一只6.5岁雄性、绝育且室内家养短毛猫（DSH），经私人诊所转诊后到兽医教学医院接受临床营养服务，该私人诊所曾为这只猫进行过牙齿清洁。这只猫在进行牙齿清洁后，被转介到服务机构，因为它被诊断为肥胖且主人愿意为自己的宠物尝试减肥计划。宠物主告诉临床医生，这只猫常常久卧不动，家里也没有其他的宠物。每天给猫喂食一种室内猫维持需要用的干粮供自由采食，还提供一点牛奶、一些奶酪零食、鸡肉或是金枪鱼，主人也不知道具体数量。隔夜禁食后，参考兽医血液学和生物化学指标显示，除了表现高胆固醇血症（含量为7.54 mmol/L，参考值为2.46~3.37 mmol/L）外，其他指标正常。

9.4.2 病情评估

在体格检查期间，猫的生命体征正常，除了肋骨上多余的脂肪，没有明显的腰部，以及腹股沟脂肪垫很明显外，其他没有发现异常。体重（BW）为10.8 kg，体况评分（BCS）被确定为9/9[1]。患宠的问题清单包括肥胖症和高胆固醇血症。

9.4.3 治疗方案

减肥计划是根据2014年美国动物医院协会（AAHA）的犬猫体重管理指南[2]计算的。估计体况评分（BCS）9/9，体内脂肪含量为45%。基于此，计算出理想体重。

理想体重计算：

10.8 kg × (100% − 45%)/0.8 = 7.4 kg

对于严重超重的猫，重要的是要采用缓慢的减肥计划，以减少猫患脂肪肝的风险[3]。应密切监视该猫，如果该猫停止进食，宠物主应按指示做出反应。

每日能量需求为静息能量需要（RER）的80%，预计每周体重减少0.5%~2%，通常被认为是安全的体重减轻率[2]。

静息能量需要（RER）的计算：

70 × 理想体重$^{0.75}$ = 70 × 7.4$^{0.75}$ = 314 kcal/d

减肥的每日能量分配计算：

0.8 × 314 = 251 kcal/d

选择的商业减肥粮的能量分布如表9.2所示。

表 9.2 减肥粮的能量分布

营养	饲喂量	代谢能
蛋白质 /%	34	42
脂肪 /%	9	24
碳水化合物 /%	28	34
代谢能(ME)/(kcal/kg)	3 090	

计算喂食量：

251/3 090 × 1 000 g/kg = 81 g/d(分为 2 等份)

美国国家研究委员会(NRC)建议的猫蛋白质需要量：

4.96 g 蛋白质 × 理想体重$^{0.67}$ = 4.96 × 7.4$^{0.67}$ = 19 g/d

建议饲喂食物的蛋白质水平计算：

0.34 × 81 g = 27.5 g/d

宠物主被告知只能喂推荐的食物，如果她想给宠物喂零食，需要少给一些干粮。

9.4.4 短期随访

该猫在 14 d 后复检。预期的体重减少量约为 200 g。但是这只猫只减少了 20 g。宠物主人一直按照规定喂食，尽管该猫一直有乞食行为，但她严格按照建议的剂量饲喂了 2 周时间。该猫也没有从其他地方获取食物。因为我们刚刚开始招募临床减肥试验，所以我们问主人是否愿意将她的猫纳入其中。她接受了建议，于是该猫开始了临床减肥试验。开始试验时，分析其血液样本和尿液样本，通过全血细胞计数和血清生物化学分析，除了发现空腹高胆固醇血症(胆固醇含量为 8.66 mmol/L，参考值为 2.46~3.37 mmol/L)外，总甲状腺素(T4)含量或尿液分析中均未发现异常。随后，使用双能 X 线吸收法(DEXA,图 9.4)分析机体组成，并使用此方法确定体脂率(BF)为 60.2%。基于新的测量结果，根据 AAHA 减肥指南[2]计算获得了新的减肥计划。

计算 BF 为 60% 时的理想体重：

现在 BW × (100 − BF)/0.8 = 10.8 × (100% − 60%)/0.8 = 5.4 kg

计算静息能量需要(RER)：

70 × 5.4$^{0.75}$ = 248 kcal/d

减肥时的每日能量分配计算：

0.8 × 248 kcal = 198 kcal /d

该猫喜欢食用这款减肥处方粮，因此决定继续喂这种食品。

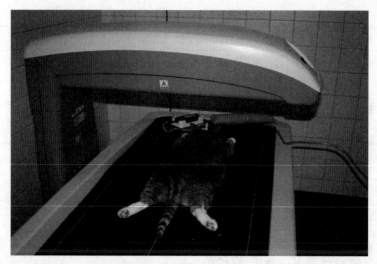

图 9.4 双能 X 线吸光扫描仪对猫进行扫描（照片由 K.M. Hoelmkjaer 拍摄）

计算喂食量：

198/3 090 × 1 000 g/kg = 64 g/d（分为 2 等份）

NRC 建议为猫提供蛋白质需要量：

4.96 g 蛋白质 × 理想 $BW^{0.67}$ = 4.96 × $5.4^{0.67}$ = 15.4 g/d

建议饲喂食物的蛋白质水平计算：

0.34 × 64 g/d = 21.8 g/d

9.4.5 讨论

根据实际的体脂率（BF），发现在过去的 14 d 中，给该猫饲喂等于理想体重的 RER 的食物，并没有被限制在预期能减肥的水平上。将 BF 调整为 60% 后，猫开始以每周 1% 的速度减轻体重。BCS 系统非常适合确定动物是否体重过轻、瘦弱或肥胖，但是在根据 BCS 估算脂肪率（BF）时应格外小心。与 BCS 有关的 BF 可能取决于年龄、品种和生活方式，因此在猫和犬的个体之间可能有所不同[1,5,6]。该病例强调了当动物肥胖程度超过 BCS 量表时（分数 9/9 等于 45% 体脂）评估动物的 BF 产生的特殊问题，并突出了对极度超重的动物使用 9 分制身体状况评分的局限性。为了提高确定特定个体能量限制的准确性，应尽可能增加不同的补充方法。如果无法进行 DEXA 扫描，则应在动物体脂指数（BFI）评分系统中补充 BCS 为 7~9/9 的动物的检查[7]。猫 BFI 旨在确定比 BCS 更高的 BF，其体内脂肪含量限制为 65%（BFI 70）。它是一种相对较新的方法，仅被验证过一次，因此尚不确定其在提高 BF 估计值方面的有效性，但它可作为评价肥胖患宠的一种有价值的补充方法。另外，如果有可能，减肥计划开始前给猫计算能量分配可以帮助对 RER 的估算和所需的能量限制。在开始减肥计划时,需要经常复查,因为这使您能够检测出当前建议的减肥效果是否仍未达到,并且可以刺激宠物主遵守计划。频繁拜访兽医可能会使某些猫应激很大。对于那些猫,可以进行一些远距离的复查,即主人在家里称量猫的体重,然后进行电话咨询。同样,要考虑到未达到预期的结果可能会有很多原因,包括对理想体重的估计不正确,宠物主不完全遵守计划,在其他地方获得食物或这些因素的结合。因此,请务必彻底调查导致结果不足的原因。

学习要点

(1)对于体况评分(BCS)为8~9/9的动物,应补充体脂指数(BFI)检查,该检查旨在确定较高的BF,并在可能的情况下计算当前的每日能量摄入量。

(2)经常复查以确定当前建议的减肥效果是否未达到。

(3)考虑到未达到预期结果可能有很多原因,包括对理想体重的估计不正确,宠物主不完全遵守计划以及在其他地方获得食物。

参考文献

1. Laflamme D. Development and validation of a body condition score system for cats: A clinical tool. *Feline Practice* 1997;25(5/6):13–18.
2. Brooks D, Churchill J, Fein K et al. 2014 AAHA weight management guidelines for dogs and cats. *Journal of the American Animal Hospital Association* 2014;50(1):1–11.
3. Armstrong PJ, Blanchard G. Hepatic lipidosis in cats. *Veterinary Clinics Small Animals* 2009;39:599–616.
4. National Research Council of the National Academies. Nutrient requirements and dietary nutrient concentrations. In: *Nutrient Requirements of Dogs and Cats*. Washington, DC: National Academies Press; 2006, pp. 354–370.
5. Bjornvad CR, Nielsen DH, Armstrong PJ et al. Evaluation of a nine-point body condition scoring system in physically inactive pet cats. *American Journal of Veterinary Research* 2011;72(4):433–437.
6. Jeusette I, Greco D, Aquino F et al. Effect of breed on body composition and comparison between various methods to estimate body composition in dogs. *Research in Veterinary Science* 2010;88(2):227–232.
7. Witzel AL, Kirk CA, Henry GA, Toll PW, Brejda JJ, Paetau-Robinson I. Use of a morphometric method and body fat index system for estimation of body composition in overweight and obese cats. *Journal of the American Veterinary Medical Association* 2014;244(11):1285–1290.

9.5 病例5：家养短毛猫减肥对糖尿病的缓解作用
ANDREW MCGLINCHEY AND MARTHA G. CLINE

9.5.1 既往病史

患猫为一只13岁的、已绝育的雌性家养短毛猫，她最初的护理兽医建议对其糖尿病(DM)和肥胖症进行营养管理，之后患猫被送往一家专科医院的临床营养服务中心。就诊的4个月前，在对多尿和多饮(PU/PD)症状进行评估后，它被诊断为糖尿病。确诊时的验血结果显示有轻度的高白蛋白血症(4.1 g/dL，参考范围2.5~3.9 g/dL)，高血糖症(412 mg/dL，参考范围64~170 mg/dL)和高甘油三酸酯血症(304 mg/dL，参考范围25~160 mg/dL)，其余参数在正常范围内。尿检结果显示糖尿症阳性、酮尿症阴性。到营养服务中心就诊时，她每12 h使用2个单位的甘精胰岛素(100 单位/mL)[①]。宠物主人在家里的多次测试结果显示糖尿症阳性、酮尿症呈阳性。确诊3个月后的果糖胺检测结果(350 μmol/L，参考范围142~450 μmol/L)与对糖尿病情的良好控制保持一致。由于患猫性格暴躁，未能进行血糖曲线测定。当时它的多尿和多饮症已有所消退，体重稳定。

9.5.2 患猫评估

就诊时，患猫每天早晚各喂1次，每次为1/4杯(8盎司量杯)市售针对糖尿病治疗的猫膨化干粮[②]，外加1茶匙同种罐装湿粮[③]。每天2次市售的用于关节保健的猫零食[④]。每日总热量消耗量估计约为319 kcal。

初步的体检显示患猫有中度牙结石、牙龈充血和肥胖症。体重为8.1 kg，体况评分(BCS)为9/9[1]，估计体脂率在50%，肌肉状况良好。其他体检项目没有特殊发现，就诊时的主要问题包括糖尿病、肥胖症，以及牙周疾病。

9.5.3 治疗方案

为了解决患猫的肥胖问题并提升胰岛素敏感性，建议开始实行减肥计划。根据她的初始体重和50%的体脂率，评估它的理想体重在5.4 kg左右。

理想体重计算 = 目前体重 × [(100 − 目前体脂%)/(100 − 理想体脂%)]

---

① 来得时胰岛素，赛诺菲公司生产。
② 普瑞纳冠能处方粮糖尿病饮食管理®猫配方干粮(592 kcal/杯，4 118 kcal/kg，蛋白质50% ME，脂肪37% ME，碳水化合物13% ME)。
③ 普瑞纳冠能处方粮糖尿病饮食管理®猫配方罐头湿粮(191 kcal/罐，1 227kcal/kg，蛋白质38.8% ME，脂肪58% ME，碳水化合物3.3% ME)。
④ 健安喜Ultra Mega关节健康护理零食(5.4 kcal/块)。

8.1 kg × (100% − 50%)/(100% − 25%) = 5.4 kg

鉴于患猫的年龄较高,以及体况评分为 5/9 的情况下雌猫体脂百分比相对更高[3],计算时选择的理想体脂百分比为 25%。一般对于身体状况评分在 5/9 的猫犬,建议的体脂百分比一般为 20%~24%[2]。根据理想体重计算出的静息能量需要(RER)为 250 kcal/d。

静息能量需要 =(体重 kg)$^{0.75}$ × 70

5.4kg$^{0.75}$ × 70=248 kcal/d

为了能以较为安全的每周 0.5%~2% 减肥率实现减肥目标,使用生命周期系数 0.8 计算其维持能量需要(MER)。

维持能量需要 = 生命周期系数 × 静息能量需要

0.8 × 248=198 kcal/d

建议主人按其热量需求,选择其中一种方式饲喂之前推荐的猫粮(见表 9.3),并在饭后 30 min 进行处方胰岛素给药。针对关节健康的市售零食也被纳入节食计划,每 12 h 喂 1 块。建议主人每天监测患猫血糖,并要注意由于体重减轻后胰岛素敏感性升高引起的低血糖的征兆。建议在 2 周后进行复诊,以确保安全有效的减肥率。由于患猫的暴躁性格,无法在医院或家中完成血糖曲线测量,因此也建议在 2 周后通过测量血清果糖胺,重新评估糖尿病的控制。

表 9.3　向宠主提供的 2 个饲喂选项

每日需求	源自罐头湿粮 /(kcal/d)	源自干粮 /(kcal/d)	源自零食 /(kcal/d)	总热量 /(kcal/d)
选项 1	191	0	10.8	201.8
选项 2	127	66	10.8	203.8

注:选项 1 为每 12 h ½ 罐头湿粮,选项 2 为每 12 h 1/3 罐湿粮和 8 g 干粮。

9.5.4　短期和长期随访

开始减肥计划 2 周后,为患猫复查体况。复查时患猫体重为 7.9 kg,每周减肥率为 1.2%。减肥计划 4 周后,患猫由于在家中反复出现糖尿阴性因而到兽医处就诊。当时的血糖值为 98 mg/dL(参考范围 64~170 mg/dL)。体重为 7.7 kg,说明保持较好的 1.2% 每周减肥率,体重减轻了 0.4 kg,占总体重的 4.9%。推测是胰岛素的作用缓解了糖尿病症状,所以停止了胰岛素用药。减肥计划保持不变,并建议主人在胰岛素停药后继续监测血糖值。

之后,患猫每 4 周复查一次体重。每次都重新计算维持能量需要(MER),并对喂食量进行调整,以确保安全的体重减轻率。开始减肥计划后的 15 个月后,患猫达到了 5.4 kg 的目标体重,体重减轻了 2.7 kg,占总体重 31%。在整个减肥计划过程中，糖尿病症状逐渐得以缓解。

9.5.5 讨论

猫的糖尿病通常是由于过度肥胖和β细胞功能异常引发外周胰岛素抵抗而引起的[4]。当猫的体重超过瘦体重的10%，就会产生胰岛素抵抗[5]。肥胖猫患糖尿病的概率是具有理想体况评分猫的3.9倍[6]。体重每增加1 kg，胰岛素敏感性就会降低30%，减肥后会恢复正常[7]。对患有糖尿病的超重/肥胖猫来说，减肥是首要任务，有助于在药物治疗配合下持续性缓解糖尿病症状。对于病情比较严重的猫，体重控制计划应适当延迟至病情稳定时再开展。

美国东南部的一名兽医进行的一项调查显示，在诊断出糖尿病时，97%（87/90）的兽医会建议对宠物进行饮食管理，93%的兽医会建议喂食标有低碳水化合物（LC）的猫粮[8]。此项调查数据并未包含建议宠物减肥的兽医数量。碳水化合物含量接近或低于35 g/1 000 kcal[9,13]的猫粮对糖尿病猫益处可能更大。与中等碳水化合物-高纤维含量猫粮相比，饲喂低碳水化合物-低纤维猫粮猫咪的糖尿病症状更易改善，且缓解率达到68%[10]，而前者的缓解率仅有41%。餐后高血糖会导致β细胞毒性增强并加剧糖尿病，值得注意的是，在一些超重猫中，餐后高血糖水平和持续时间都会增加[11]。因此，正如本病例所示，治疗肥胖对于缓解糖尿病情具有重要意义。仅改用低碳水化合物的猫粮而不解决超重问题，并不能起到缓解糖尿病的作用。作者通常会建议给患有糖尿病的猫饲喂一种罐装的、低碳水化合物、具有治疗作用的减肥猫处方粮，与低碳水化合物猫干粮相比，这种罐装粮的含水量更高，热量密度更低，并且方便控制分量。

一些患猫和主人可能更喜欢猫干粮。在这种情况下，建议给患猫饲喂一种低密度、高蛋白质（≥ 115 g/1 000 kcal）的减肥猫粮，以适应主人和患猫对于较大喂食量的需求。这些猫干粮可能具有较低脂肪含量和/或补充纤维，以降低热量密度。在糖尿病患猫的减肥计划中，建议饲喂高蛋白质（≥ 115 g/1 000 kcal）猫粮，以支持对瘦体重的维持。在减肥过程中，建议控制喂食量，而不是随意喂食猫干粮。可以在喂食时用厨房秤对猫干粮用量进行称重，以提高准确性[12]。具体重量可以参考宠物食品标签上提供的热量信息（kcal/kg）进行计算。

> **学习要点**
>
> （1）在开始减肥计划前，即使通过药物治疗配合低碳水化合物饮食，患猫的糖尿病症状并没有缓解。减肥对于改善超重/肥胖猫外周胰岛素敏感性至关重要。
>
> （2）在糖尿病患猫病情稳定情况下，以及整个治疗过程中，兽医和宠物主人应要能够及时辨别并设法解决肥胖症。

(3) 让主人更多地了解超重/肥胖对糖尿病控制的影响,与提供喂食建议一样重要。

(4) 低碳水化合物或高纤维的高蛋白质猫粮可用作糖尿病患者猫减肥计划的一部分。有证据表明,饲喂低碳水化合物、高蛋白质的食物有助于减少胰岛素需求,但对于不同的患猫,应提供个性化的喂食建议。

参考文献

1. Laflamme D. Development and validation of a body condition score system for cats: A clinical tool. *Feline Practice* 1997;25:13–17.
2. Brooks D, Churchill J, Fein K, Linder D, Michel KE, Tudor K, Ward E, Witzel A. 2014 AAHA weight management guidelines for dogs and cats. *Journal of the American Animal Hospital Association* 2014;50:1–11.
3. Bjornvad CR, Nielsen DH, Armstrong PJ, McEvoy F, Hoelmkjaer KM, Jensen KS, Pedersen GF, Kristensen AT. Evaluation of a nine-point body condition scoring system in physically inactive pet cats. *American Journal of Veterinary Research* 2011;72:433–437.
4. Rand JS, Fleeman LM, Farrow HA, Appleton DJ, Lederer R. Canine and feline diabetes mellitus: Nature or nurture? *Journal of Nutrition* 2004;134:2072S–2080S.
5. Hoenig M, Pach N, Thomaseth K, Le A, Schaeffer D, Ferguson DC. Cats differ from other species in their cytokine and antioxidant enzyme response when developing obesity. *Obesity (Silver Spring)* 2003;21:E407–E414.
6. Scarlett JM, Donoghue S. Associations between body condition and disease in cats. *Journal of the American Veterinary Medical Association* 1998;212:1725–1731.
7. Hoenig M, Thomaseth K, Waldron M, Ferguson DC. Insulin sensitivity, fat distribution, and adipocytokine response to different diets in lean and obese cats before and after weight loss. *American Journal of Physiology: Regulatory, Integrative and Comparative Physiology* 2007;292:R227–R234.
8. Smith JR, Vrono Z, Rapoport GS, Turek MM, Creevy KE. A survey of Southeastern United States veterinarians' preferences for managing cats with diabetes mellitus. *Journal of Feline Medicine and Surgery* 2012;14:716–722.
9. Behrend E, Holford A, Lathan P, Rucinsky R, Schulman R. 2018 AAHA diabetes management guidelines for dogs and cats. *Journal of the American Animal Hospital Association* 2018;54:1–21.
10. Bennett N, Greco DS, Peterson ME, Kirk C, Mathes M, Fettman MJ. Comparison of a low carbohydrate-low fiber diet and a moderate carbohydrate-high fiber diet in the management of feline diabetes mellitus. *Journal of Feline Medicine and Surgery* 2006;8:73–84.

11. Coradini M, Rand JS, Morton JM, Rawlings JM. Effects of two commercially available feline diets on glucose and insulin concentrations, insulin sensitivity and energetic efficiency of weight gain. *British Journal of Nutrition* 2011;106(Suppl 1):S64–S77.
12. German AJ, Holden SL, Mason SL, Bryner C, Bouldoires C, Morris PJ, Deboise M, Biourge V. Imprecision when using measuring cups to weigh out extruded dry kibbled food. *Journal of Animal Physiology and Animal Nutrition* 2011;95:368–373.
13. Bonagura JD, Twedt DC. *Kirk's Current Veterinary Therapy (CVT) XIV*. 14th ed. Elsevier, St. Louis, MO 2008:201.

9.6　病例 6：肥胖症的行为管理
DEBORAH E. LINDER 和 MEGAN K. MUELLER

9.6.1　患犬病史

Jacob 是一只 2 岁的超重已绝育雄性拉布拉多寻回猎犬，最近由于前交叉韧带撕裂，经历了侧边缝合修复手术后，进行了体重管理咨询。最初的饮食史显示，它每天从非处方减肥犬干粮摄入 600 kcal，除此之外，根据主人家的晚餐情况，它从主人剩余晚餐中摄入额外未知的热量。每天拌着奶酪和熟食肉喂抗炎药 2 次，额外增加 100 kcal。

9.6.2　患犬评估

Jacob 重 44 kg(97 lb)，体况评分为 8/9，肌肉状况正常。除了在拉伸四肢时有轻度到中度疼痛外，体检和实验室检查(包括完整的血细胞计数、电解质检测、尿液分析和甲状腺素水平)都没有明显异常，Jacob 的理想体重评估为 34 kg(74 lb)。这个理想体重是通过参考 Jacob 以往病历中，体况评分 BCS 在 5/9 时的体重确定的[1]。

9.6.3　治疗方案

手术后的最初 8 周 Jacob 被限制运动。所以在此期间，它只包括饮食管理。考虑到 Jacob 需要限制热量摄入而且由于它不能运动，主人给他换成了减肥用处方粮。最初的热量摄入量根据理想体重(34 kg)的静息能量需要计算[1]，计算公式为：$70 \times 体重_{kg}^{0.75} = 70 \times 34^{0.75} \approx 1\,000 \text{ kcal/d}$。

根据 Jacob 的情况制定每天饮食计划如下：每天热量摄入总量为 1 000 kcal，其中包括 900 kcal 的减肥用处方粮，100 kcal 包含在 Jacob 喜欢的零食中。

9.6.4 短期和长期随访

每周给 Jacob 称体重,并通过调整摄入热量以达到目标减肥率(每周 0.5%~1%)[1]。体重管理实行 2 周后复诊时,Jacob 近期不安的表现引起了关注,比如试图翻垃圾桶并毁坏主人的衣服,只有喂食的时候 Jacob 会表现出短暂的兴奋,据主人说大概只持续 10 s 的时间。

即使喂它的食物大部分都是处方粮,他还是吃的又快又高兴,但随后又变得焦躁不安,直到下一次喂食时。通过深入谈话得知,宠物主人对于 Jacob 被限制不能在他们房子附近的池塘散步或者玩耍感到十分内疚。宠物主人不想 Jacob 讨厌他或厌烦他们在一起的时光。他认为吃东西是 Jacob 生活中唯一的乐趣,但太过短暂,因此他想在 Jacob 解除运动限制之前,给它多喂些食物和零食。

我们告诉宠物主人,限制运动对于恢复身体非常必要,吃更多食物或零食会增加体重,超重会导致疼痛加剧以及情绪障碍,这会使 Jacob 感觉更糟糕[2]。美国动物医院协会(AAHA)2014 年的体重管理指南提供了针对行为管理的问题排除提示[1],其中包括使用能延长进餐时间的喂食玩具和漏食球。Jacob 的主人非常喜欢把喂食变成互动游戏的建议,这可以让 Jacob 更愉快,也能让主人享受有宠物陪伴的时光。由于每日一餐,宠主用干粮颗粒训练 Jacob,以延长喂食时间。例如,让 Jacob 在两个爪子上各放一粒犬粮时保持不动,等待主人的信号,如图 9.5 所示。这些技巧解决了主人的担忧,在第 8 周复查时,Jacob 的体重减到 35 kg(77 lb),其家庭成员则继续遵循该减重计划。

9.6.5 讨论

Jacob 能够开心并且享受与主人在一起的时光是主人与 Jacob 关系融洽的重要方面。为了让主人和 Jacob 都能从积极的互动中受益,我们鼓励以 Jacob 的进食动力作为探索创新方式的一次机会,在喂食时融入精神激励和游戏。主人没有让 Jacob 迅速吃完所有食

图 9.5 每日一餐,主人用干粮颗粒教 Jacob 新的花招,延长了喂食时间。在这张照片中,主人教它待着不动,在它的每只爪子上放一粒犬粮,然后等着吃

物,而是让它接受了积极的训练,这能够增进他们之间的感情,同时又不会产生负面影响。主人看得出来,Jacob喜欢学习新技巧,而且Jacob在兴奋地找到自己食物的同时,与他共度了美好的时光。

> **学习要点**
>
> (1)肥胖症是一种复杂的营养失调疾病,通常需要进行综合管理,而并不只是标准食谱和运动计划就能成功减肥。
>
> (2)有效的客户沟通可以帮助兽医团队了解每个家庭与宠物的独特关系,并保持人与动物之间的纽带不受破坏。
>
> (3)肥胖症治疗可以有多种模式,从医学到心理学的多个方面进行管理才能够成功实现行为管理。
>
> (4)宠物主人可以与兽医一起制定,这样可以加强主人与宠物的关系,并且不会影响他们的亲密关系。

参考文献

1. Brooks D, Churchill J, Fein K, Linder D, Michel KE, Tudor K, Ward E, Witzel A. 2014 AAHA weight management guidelines for dogs and cats. *Journal of the American Animal Hospital Association* 2014;50:1–11.
2. German AJ, Holden SL, Wiseman-Orr ML, Reid J, Nolan AM, Biourge V, Morris PJ, Scott EM. Quality of life is reduced in obese dogs but improves after successful weight loss. *Veterinary Journal* 2012;19:428–434.

9.7 案例7:一例新型蛋白质饮食试验中的热量限制、运动和治疗性减肥
JUSTIN SHMALBERG

9.7.1 既往病史

一只2岁的绝育雄性杂交大型犬到一位经委员会认证的皮肤科医生那就诊,该犬从3个月大开始就被发现存在非季节性的皮肤瘙痒问题。宠物主人描述瘙痒是弥散性的,并间歇性发生皮肤病变,最初出现在爪子上,之后逐渐蔓延到腹部、腹股沟、腋窝和耳朵区域,并常伴有继发性细菌感染和马拉色菌感染。血清检测发现了多种树木、禾本科草类、杂草、霉菌、昆虫和室内过敏原的IgE抗体。一些针对食

物抗原的抗体也有发现,包括鸡蛋、燕麦、土豆、大米和大豆。患犬以前的治疗包括头孢氨苄、酮康唑和羟嗪的反复给药,基于血清 IgE 结果的多种经验性饮食试验方案和免疫注射治疗,以及间断性的给予逐渐减量的糖皮质激素。抗真菌或抗生素治疗能观察到部分治疗效果,而持续 18 个月的免疫治疗方案并没有改善临床症状。主人表示在患犬骨骼成熟后,其体重一直在逐渐增加。

9.7.2 患犬评估

该患犬体况评分(BCS)为 9/9,具有肥胖症,体重为 62.7 kg,肌肉情况正常(肌肉状况评分为 3/3)。患犬的生命体征正常,心胸听诊无显著异常。患犬具有弥漫性瘙痒,腹部有脓疱、丘疹和表皮环形脱屑,在每个爪的指间隙中出现轻度红斑,存在双侧表皮结痂和多灶性躯干脱毛。在对右后肢进行股骨伸展检测时,患犬表现抗拒,并表现出该关节的活动范围下降。其他无异常发现。

患犬的皮肤病变与继发性细菌性脓皮病和马拉色菌皮肤炎一致。可以通过过敏性皮炎、跳蚤过敏皮炎、接触性过敏或食物过敏等方面对诱发病因进行区别诊断。皮肤科医生开了甲氧苄啶-磺胺嘧啶(TMS)、局部曲安奈德、酮康唑和医用沐浴液。建议通过营养咨询,对当前饮食摄入量、食物减量试验的可能性进行评估,以达到减肥的目的。

主人主要喂食一种市售的以兔肉为主的无谷物犬干粮①,估计每日提供的热量约为 1 928 kcal。同时在喂的还有干鸭肉类零食②,平均每天喂 3 个零食(57 kcal)。其他补充性食物没有提及。患犬之前的食物接触史包括一些种类的市售食品,通过每 2~3 个月更换种类来评估效果。对这些产品的调查显示,患犬食用过牛肉、野牛肉、羔羊肉、鹿肉、白鱼、鸡蛋、金枪鱼,以及多种蔬菜和谷物。他们咨询的兽医推荐根据选择兔肉粮进行新的蛋白质饮食试验,因为列表上的成分都经过了 IgE 测试,并且据宣传"可能缓解大多数食物过敏症状"。然而,该饮食除了有兔肉外还包含鲑鱼、猪肉制品和植物蛋白质,因此用来开展饮食试验并不合适。

患犬肥胖是由于缺乏运动和绝育导致其瘦体重下降[1],而长期以来摄入的热量超过他自身需求。例如甲状腺功能减退造成代谢率的改变,进而更易肥胖的各种病理状态都可能是诱发肥胖的潜在原因。然而,该犬开始肥胖时的年龄使得无法对病因加以鉴别。糖皮质激素治疗期间提供超过预期热量需求的高能量食物可能与病史和临床表现更为一致。患犬不常运动,因此并没有比那些长期不动的犬消耗更多的热量。在就诊前的 3 个月中,患犬体重稳定,根据身体状况评分估算理想体重是 42 kg,而其超重约 50%(每超过 5/9 1 分,对应超出理想体重的 13%)[2]。

---

① 百利本能兔肉配方,482 kcal/杯,3 988 kcal/kg。
② Dogswell 关节保健鸭胸零食,19 kcal/个零食,每千克中含的热量信息未知(该产品停产)。

患犬的热量摄取（121× 理想 $BW_{kg}^{0.75}$）与同龄实验犬接近,实验犬的热量摄取通常比宠物犬高 30%[1]。大多数报告中都提到犬的维持能量需要（MER）差异比较大,但很明确的是为减轻体重,必须大量减少热量摄入。

9.7.3 治疗方案

相比小型犬或体重较轻的犬,大型犬和肥胖犬需要更大的热量限制才能达到推荐的每周 1% 的减肥率[4],因此,将患犬的静息能量需要（RER,70× 理想 $BW_{kg}^{0.75}$）约 1 150 kcal 定为最初目标。目标为 RER 的能量摄入应该是足够的,因为与宠物犬相比,患犬体重稳定时的热量摄入要更高一些。自制犬粮可能是排除试验的最佳选择。

在最初的食物试验中,为了简化步骤和避免混淆添加剂原料,不均衡的自制食物曾经备受推崇[5,6]。但患犬要进行热量限制,上述方法存在一定缺陷。因此,制定了由鳄鱼尾肉和藜麦组成的配方粮,以满足成年患犬在理想体重下的推荐摄入量（RA）（表 9.4）。通过一些资料确定了鳄鱼尾的基本营养,并从未发表的数据中获得了其氨基酸组成。关于矿物质和维生素的分析结果不清楚。所有添加剂都对潜在抗原进行了评估,唯一可能的过敏源是复合维生素中的胭脂红（一种天然着色剂）,但并不确定患犬之前是否有过接触史。配方粮中的大部分营养成分达到或超过了患犬的推荐摄入量 RA（表 9.5）,而对于那些低于推荐摄入量 RA 的营养成分,无法确定鳄鱼肉中这些成分的信息,推测这些成分在配方粮中并不缺乏。

表 9.4　新型蛋白质减肥粮配方（每日平均量）

成分	用量
鳄鱼尾肉	330 g
干藜麦	150 g
葵花籽油,65% 亚油酸	2 汤匙
亚麻籽油	2 茶匙
鱼油	1 茶匙
碘盐	1/4 茶匙
低钠盐	1/2 茶匙
磷酸氢钙	1.5 茶匙
配方 2（复合维生素）[a]	1 片
酒石酸氢胆碱,300 mg 胆碱/片[b]	2 片
L-蛋氨酸粉,2.8 g/匙[c]	1/4 茶匙

[a] Integrative Therapeutics, Green Bay, Wisconsin.
[b] Country Life, Hauppauge, New York.
[c] Life Extension, Fort Lauderdale, Florida.

表 9.5 所选食物营养素与 2006 年 NRC 推荐摄入量(RA)的比较

营养成分	单位	自制粮			摄入量
		42 kg 重的犬	每天摄入量	RA/%	
能量	kcal	2 145	1 280	60	1 000
粗蛋白质	g	53.6	102	191	79.8
脂肪	g	29.6	59.5	201	46.5
亚油酸	g	6.0	22	372	18
ALA	g	0.2	5.0	2 132	3.9
EPA 和 DHA	g	0.2	1.8	780	1.4
钙	g	2.1	2.1	97	1.6
磷	g	1.6	2.0	124	1.6
维生素 A	RE	813	3 000	369	2 340
胆钙化醇	μg	7.3	10	137	7.8
维生素 E	mg	16	60	372	47
硫胺素	mg	1.2	15	1 274	12
核黄素	mg	2.8	16	559	12
吡哆醇	mg	0.8	25.3	3 149	19.8
烟酸	mg	9.1	29.4	323	23.0
泛酸	mg	8.0	26.6	330	20.8
钴胺素	μg	18.8	25.0	133	19.5

通过配方粮摄入的热量约为 1 280 kcal/d,略高于既定目标。但比以前的热量摄入减少了 1/3 以上,第一次复查时会对配方粮进行调整。客户拒绝了商业实验室对配方粮营养成分分析的服务。建议每隔 3~5 周进行一次复查,但因为患犬同时在进行皮肤病的治疗,皮肤科医生的诊所距离较远,因此最终的复查时间取决于皮肤科复查的时间。在患犬的饮食随访中验血不是必需的,一是因为患犬年龄比较年轻,二是自制粮只用来进行短期喂养,即在皮肤感染减轻后和口服食物激发试验开始前的最多 4~6 周。总体营养预后并不确定,这是因为只有通过对之前食物或单一蛋白质的过敏性检测,才能确定食物的过敏性[5],并且在其他类型过敏性皮肤疾病的患犬中,对于脂肪酸添加剂的反应也是无法预测的[7]。

由于患犬和其主人都没有运动的习惯,进行运动减肥有一定的难度。尽管肥胖会加剧骨骼肌疾病的发展和严重程度,但患犬轻度的跛脚并不需要特别的治疗。运动过少可能会影响每日能量消耗。因此对主人提出了遛犬的建议,并希望在 1 个月后能达到每天 2 次,每次 30 min 的运动量。按 1.6 km/h 的步行速度和 1.2 kcal/(kg$^{0.75}$·km) 的能量消耗计算,达到运动目标后的日均热量消耗约为 30 kcal[9]。水下跑步机的运动方式可能更有助于保持瘦体重[10],但与陆地训练相比,每天只增加 18 kcal 的消耗,而且还会增加成本[9]。最终主人选择在家锻炼的方案。

9.7.4 随访

通过吃自制粮,最初 3 周患犬的平均周减肥率达到初始体重的 1.6%,未检测到肌肉质量下降(表 9.6)。在此期间,所有的经调味的口服药物均已停药。

表 9.6 监测期间体重史和体况评分

时间 / 周	0	3	11	14	19	23	29	35
体重 /kg	62.7	59.6	54.8	52.8	50.4	47.9	44.9	42.3
每周减肥率 /%[a]	—	1.6	1.0	1.2	0.9	1.2	1.0	1.0
体况评分(BCS)	9	9	8	8	7	7	6	5

[a] 每周减肥率的计算基于之前复查时体重的平均值。用初始体重计算的平均减肥率是 0.93%。

患犬的皮肤瘙痒有所缓解,皮肤细胞学检查呈球菌阴性,之前的药物治疗仍在继续。初期减肥率往往比后期要高[11],所以并没有调整热量摄入量。患犬的状况在抗生素疗程 4 周后有所好转,瘙痒程度在 10 分量表上从 9 分改善为 2 分。比较遗憾的是,由于短吻鳄肉供应不上,尝试将其换成罗非鱼作为新的蛋白质来源。结果患犬的瘙痒评分在 48 h 内回升到 8~9 分。之后主人通过其他渠道获得了鳄鱼尾,更换成之前的配方,在接下来的 1 周瘙痒评分又恢复到 4 分。

第 11 周进行复查时显示,自第 3 周以来,每周的平均减肥率是 1%,同时体况评分 BCS 降到 8 分。主人也一直按计划对宠物进行锻炼。细菌性皮炎持续存在,皮内试验显示对尘螨、马拉色菌、几种草和植物产生轻度至中度的反应,这与观察到的临床症状并不一致。第 14 周患犬症状部分缓解,减肥率还是每周 1%,对自制粮适应较好。细菌培养结果显示对 TMS 具有耐药性而对氯霉素敏感,因此开始用氯霉素进行治疗。治疗 8 周后,患犬的临床症状逐渐恶化,考虑到阳性反应以及还可能存在的食物过敏,并没有进行进一步的接触性或皮内过敏试验。皮肤科医生建议 6 周内进行复查。复查前 3 天,在没有改变饮食情况下,患犬复发皮肤瘙痒。当时的体况评分(BCS)是 6/9,每周平均减肥率是 1%,无明显肌肉萎缩。空气过敏源的皮内试验结果与过敏性皮炎一致。皮肤科医生建议在过敏病症得到控制之前,继续采用之前的自制粮。2 个月后开始免疫治疗,并且换了一种非处方的以鸡肉为主的维持粮,换粮后没有引起临床症状的复发。换粮并引起食物过敏,但引发了严重的间歇性过敏性皮炎和细菌性皮炎,这使食物试验受到干扰。尽管如此,经过配方食物和运动调整后,患犬的体重得到控制,没有明显的肌肉损失,为确保体重降低到目标体重,一直对食物热量摄入进行监测,并在试验开始后的4年内一直保持。

9.7.5 讨论

控制肥胖可以采用多种饮食策略。参考之前的病例[4],一般来说,减肥食谱的配方中会增加营养成分的含量,以确保在限制热量摄入过程中营养不会缺乏。降低能量密度的方法包括增加纤维含量及降低膳食脂肪。减肥粮一般会通过限制膳食脂肪含量来降低能量密度,增加热量损耗,但由于使用特定的限定原料和为了提高多不饱和脂肪酸含量,本病例的配方粮并没有改变营养物质的用量。膳食蛋白质在减肥中的作用并不是很明确。

自制粮里较高的蛋白质含量(>26%的代谢能ME或>75 g/Mcal)能够保持去脂体重[12],并可能促进能量消耗和适口性[3,25]。在不限制热量摄入的情况下,光靠运动促进减肥似乎是不够的,与限制热量摄入配合运动的减肥方案相比,两者的效果还是存在差异的(见第7章)。

患犬从青少年时期开始的非季节性瘙痒可能与食物过敏有关,因此增加了肥胖症管理的复杂性,所以有必要在实行减肥计划前考虑多方面因素。食物过敏是一种不良的食物反应,过敏源尚不清楚,也不确定与品种、性别和年龄的相关性[5]。由IgE介导的快速过敏反应和迟发性过敏都可能是犬类食物过敏的原因[5]。通常情况下,IgA等分泌型抗体,以及由肠系膜淋巴结和肠道相关淋巴组织调节的T细胞抑制,能够阻止对食物来源抗原的异常反应[13]。但是,在黏膜屏障受损、出生后口服耐受性发育不良、微生物菌群变化或并发疾病情况下,上述保护机制可能失效[5,13]。有趣的是,受感染的犬主要表现出皮肤症状[14],且很难在临床上与空气过敏源引起的过敏性皮炎加以区分。最近的研究发现,犬类表皮中的T细胞表型可能与皮肤的食物过敏性有关[15]。目前还没有针对上述研究的诊断检测方法。由于犬类过敏性皮炎也会引起IgE水平增加[16,17],血清中IgE的含量指标也并不可靠。因此,对食物过敏的诊断需要一种新型的蛋白质饮食试验,以及详细的饮食史[5]。

由于患犬之前广泛的接触史,市售的限定原料的处方粮并不适用于新型蛋白质配方粮试验。而非处方粮没有经过具体评估,一些产品中包含未在标签注明的食物抗原[18],不适用于减肥。水解后的食物也曾在考虑范围内,但是对于特定病患,如果完整的蛋白质是过敏源的话,它的小分子水解产物也可能引发食物过敏[19]。而且,这类粮的蛋白质含量较低(ME为16.5%~18%,或47~52 g/Mcal),一方面可能是加工成本的原因,另一方面可能是由于存在食物抗原具有剂量依赖性这样的错误观念[5]。因此,市售犬粮都无法满足新型高蛋白质减肥配方粮的基本要求。

本病例的配方粮改变了多不饱和脂肪酸(PUFAs)的用量。人类和犬类过敏性疾病的过敏源常来自环境而非食物[7]，但在治疗中对 PUFAs 用量进行调整也较为常见。这种调整有助于缓解食物过敏，对一些食物和空气过敏源双重敏感的病患也有益处[5,20]。之前的关注重点一直放在增加 ω-3 脂肪酸的含量[7]，但是亚油酸对于稳定角质层屏障至关重要，而在人类过敏性疾病中限制 n-6 PUFA 的做法也遭到质疑[21]。因此，配方粮中增加了 n-3 和 n-6 PUFAs 的用量。之前的报道中对于用量的建议存在不同意见[22,23]，因此最终的用量参考了必需脂肪酸的建议用量[1]。

在本减肥病例中，饮食和运动都发挥了作用。症状和临床表现上的复杂性提示个性化定制营养策略的适用性，在一些案例中也确实有助于达成减肥目标。由于普遍存在体重反弹的情况，因此有必要对减肥患宠进行长期监测[24]。

> **学习要点**
>
> (1)大多数犬类的目标减肥率在每周 0.5%~2%，前几周的减肥率可能更高。
>
> (2)存在其他营养性反应状态下的肥胖症管理可能需要对动物的饮食计划进行个性化定制。
>
> (3)自制的食物也可以用来控制体重，前提是要精心计划，并包含一些已知的能够促进减肥成功率的因素，例如高蛋白质含量。自制食物建议咨询认证的兽医营养师。

参考文献

1. National Research Council (NRC) Ad Hoc Committee on Dog and Cat. *Nutrient Requirements of Dogs and Cats.* Washington DC: National Academies Press; 2006.
2. German AJ, Holden SL, Bissot T et al. Use of starting condition score to estimate changes in body weight and composition during weight loss in obese dogs. *Research in Veterinary Science* 2009;87(2):249–254.
3. Diez M, Nguyen P, Jeusette I et al. Weight loss in obese dogs: Evaluation of a high protein, low-carbohydrate diet. *The Journal of Nutrition* 2002;132(6):1685S–1687S.
4. Brooks D, Churchill J, Fein K et al. 2014 AAHA weight management guidelines for dogs and cats. *Journal of the American Animal Hospital Association* 2014;50:1–11.
5. Verlinden A, Hesta M, Millet S et al. Food allergy in dogs and cats: A review. *Critical Reviews in Food Science and Nutrition* 2006;46(3):259–273.
6. Roudebush P, Cowell CS. Results of a hypoallergenic diet survey of veterinarians in North America with a nutritional evaluation of homemade diet prescriptions. *Veterinary*

Dermatology 1992;3(1):23–28.
7. Mueller RS, Fieseler KV, Fettman MJ et al. Effect of omega-3 fatty acids on canine atopic dermatitis. *Journal of Small Animal Practice* 2004;45(6):293–297.
8. Marshall W, Hazelwinkel H, Mullen D et al. The effect of weight loss on lameness in obese dogs with osteoarthritis. *Veterinary Research Communications* 2010;34(3): 241–253.
9. Shmalberg J, Scott KC, Williams JM, Hill RC. Energy expenditure of dogs exercising on an underwater treadmill compared to that on a dry treadmill. *13th Annual AAVN Clinical Nutrition and Research Symposium*, Seattle, Washington; 2013.
10. Vitger AD, Stallknecht BM, Nielsen DH, Bjornvad CR. Integration of a physical training program in a weight loss plan for overweight pet dogs. *Journal of American Veterinary Medical Association* 2016;248(2):174–182.
11. Fritsch DA, Ahle NW, Jewell DE et al. A high-fiber food improves weight loss compared to a high-protein, high-fat food in pet dogs in a home setting. *The Journal of Applied Research in Veterinary Medicine* 2010;8(3):138–145.
12. Hannah SS, Laflamme DP. Increased dietary protein spares lean body mass during weight loss in dogs (ACVIM abstract). *Journal of Veterinary Internal Medicine* 1998;12:224.
13. Brandtzaeg P. Food allergy: Separating the science from the mythology. *Nature Reviews Gastroenterology and Hepatology* 2010;7:380–400.
14. Veenhof EZ, Knol EF, Schlotter YM et al. Characterisation of T cell phenotypes, cytokines and transcription factors in the skin of dogs with cutaneous adverse food reactions. *The Veterinary Journal* 2011;187(3):320–324.
15. Veenhof EZ, Rutten VP, van Noort R et al. Evaluation of T-cell activation in the duodenum of dogs with cutaneous food hypersensitivity. *American Journal of Veterinary Research* 2010;71(4):441–446.
16. Jackson HA, Jackson MW, Coblentz L et al. Evaluation of the clinical and allergen specific serum immunoglobulin E responses to oral challenge with cornstarch, corn, soy and a soy hydrolysate diet in dogs with spontaneous food allergy. *Veterinary Dermatology* 2003;14(4):181–187.
17. Foster AP, Knowles TG, Moore AH et al. Serum IgE and IgG responses to food antigens in normal and atopic dogs, and dogs with gastrointestinal disease. *Veterinary Immunology and Immunopathology* 2003;92(3–4):113–124.
18. Raditic DM, Remillard RL, Tater KC. ELISA testing for common food antigens in four dry dog foods used in dietary elimination trials. *Journal of Animal Physiology and Animal Nutrition* 2011;95(1):90–97.
19. Cave NJ. Hydrolyzed protein diets for dogs and cats. *Veterinary Clinics of North America: Small Animal Practice* 2006;36(6):1251–1268.
20. Nesbitt GH, Freeman LM, Hannah SS. Effect of n-3 fatty acid ratio and dose on clinical manifestations, plasma fatty acids and inflammatory mediators in dogs with pruritus. *Veterinary Dermatology* 2003;14(2):67–74.

21. Sala-Vila A, Miles EA, Calder PC. Fatty acid composition abnormalities in atopic disease: Evidence explored and role in the disease process examined. *Clinical & Experimental Allergy* 2008;38(9):1432–1450.
22. Bauer JE. Therapeutic use of fish oils in companion animals. *Journal of the American Veterinary Medical Association* 2011;239(11):1441–1451.
23. Lenox CE. Timely topics in nutrition: An overview of fatty acids in companion animal medicine. *Journal of the American Veterinary Medical Association* 2015;246(11):1198–1202.
24. Laflamme DP, Kuhlman G. The effect of weight loss regimen on subsequent weight maintenance in dogs. *Nutrition Research* 1995;15(7): 1019–1028.
25. Toll PW, Yamka RM, Schoenherr WD et al. *Obesity.* In Hand MS, Thatcher CD, Remillard RL et al. (eds.) Small Animal Clinical Nutrition. Topeka, Kansas: Mark Morris Institute, 2010;502–542.

9.8　病例 8：体重减轻计划中兽医护士的作用
ASHLEY COX

9.8.1　既往病史

一只 7 岁超重已绝育雌性的彭布洛克威尔士柯基犬（Sugar）因减肥方案到一所兽医教学医院的临床营养服务中心就诊。Sugar 最近刚从一次柯基犬救助中被领养，几乎没有能够参考的医疗信息。新主人非常希望能够帮助 Sugar 减肥。

9.8.2　饮食史

初次就诊前，Sugar 的主人填写了一份关于饮食、活动能力和家族史的表格（DAHHF），这个调查表是在预约就诊时由接待员通过邮件发送的。就诊时，主人需要带上这个调查表。就诊当天，兽医护士对 Sugar 进行称重，并查看了它的饮食史。被领养之前，Sugar 每天饲喂 3/4 杯由 2 种干粮等比配制的混合干粮，估计每天的热量摄入是 577 kcal。2 个月前 Sugar 被收养之后，主人把喂食量降到了每次 3/8 杯，一天喂 2 次，每天加喂 1 杯绿豆作为零食，并补充 1 块有助关节健康的保健咀嚼零食。在这 2 个月的时间里，Sugar 每天通过正餐摄入热量大概为 299 kcal（占每日摄入热量的 74%），通过零食的摄入热量为 47 kcal，通过营养补充剂摄入大概 58 kcal（每日摄入热量的 26%），总计 394 kcal/d。虽然不是很确定，但主人认为这段时间 Sugar 可能还是减掉了一些体重。

9.8.3　患犬评估

最初就诊时，Sugar 除了肥胖，体检无其他异常。兽医和兽医护士通过触诊，估计 Sugar 的体况评分（BCS）约为 9/9，体脂率为 50%~55%，肌肉状态正常，体重为 18.4 kg（40.4 lb）。基础全血细胞计数（CBC）、生化指标、总甲状腺素（T4）和尿液分析无异常。我们给 Sugar 拍摄了站立时的照片，包括俯视照、侧面照（图 9.6a）和背面照，并把这些照片添加到 Sugar 的病历中。在主人签署了照片发布许可后，这些照片被放到客户等候区中的减肥计划参与者的展示板上。

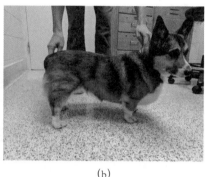

(a)　　　　　　　　　　　　(b)

图 9.6　(a)减肥计划开始前（第 0 周）和(b)减肥计划完成后（第 52 周）的侧面照

9.8.4　治疗方案

基于评估的 52.5% 体脂率，根据 2014 年 AAHA 犬猫体重管理指南，对 Sugar 的理想体重进行估算[1]。

理想体重的计算：

18.4 kg × (100% − 52.5%)/80% = 10.9 kg（24.0 lb）

每日静息能量需要（RER）约为 420 kcal[1]。

静息能量需要（RER）计算：

$RER = 70 \times 理想体重_{kg}^{0.75} = 70 \times 10.9^{0.75} = 420 \text{ kcal/d}$

减肥的每日能量分配计算：

1.0 × 420 = 420 kcal/d

0.8 × 420 = 336 kcal/d

建议每天喂食总热量 400 kcal，包括从治疗性减肥粮中摄取 360 kcal，从零食和营养补充剂中摄取 40 kcal。首次就诊时，主人可以选择以更为优惠的价格购买"减肥套餐"，其中包括了首次就诊及之后 12 个月每月 1 次的由兽医护士进行的体重复查，或者也可以选择对每次复查单独结算。Sugar 的主人购买了减肥套餐，并预约了 4 周后进行复查。离开前，主人被邀请加入由所有减肥计划参与者组成的 Facebook 私人专页①，并被建议其参加每月由诊所举办的互助会。诊所还为 Sugar

---

① Facebook 公司，加利福尼亚州门洛帕克市。

提供了一个"启动包",以激励它在整个减肥事业中坚持下来。

9.8.5 随访

客户服务代表按预约的时间电话提醒,Sugar 在 6 周后进行了第一次体重复查。复查体重为 16.7 kg(36.7 lb),减肥 1.7 kg(3.7 lb),总体重的 9.2%。主人说 Sugar 对新的减肥粮适应得很好,也一直根据减肥计划监控热量摄入。绿豆还是作为零食,每天约 2/3 杯,加 1 汤匙南瓜罐头(共计 31.6 kcal),半块关节保健营养补充剂软咀嚼颗粒(12 kcal)。每日总计摄入 404 kcal,其中 89% 来自正餐,11% 来自零食和营养补充剂。虽然来自零食的热量略高于建议的 10%,但 Sugar 的减肥率也达到了每周体重的 1.5%(框 9.1),减肥计划执行良好。之后 7 个月,Sugar 约每 4 周复查 1 次,减肥率一直保持较好。

框 9.1　每周减肥率的计算

$$总减肥率 = \frac{(先前体重 - 现在体重)}{先前体重} \times 100$$

$$Sugar\ 的总减重率 = \left[\frac{18.4\ kg - 16.7\ kg}{18.4\ kg}\right] \times 100 = 9.2\%$$

$$每周减重率 = \frac{总减重率}{自上一次减重以来的周数}$$

$$Sugar\ 每周的减重速度 = \frac{9.2\%}{6} = 1.5\%$$

---

7 个月后复查时发现,Sugar 的减肥进入平台期。对它的能量摄入的回顾中显示,主人并没有改变饮食成分及用量。因此,建议主人在喂食时用厨房秤按克称取干粮来增加精确度[2]。还和宠主讨论了使用喂食玩具和自动喂食器,以消除宠主和食物喂食行为的联系。之后主人将喂食量精确到克,配合喂食玩具,将总量降低到静息能量需要的 80%(减少 68 kcal),并参加每月的互助会和护士复查。在首次就诊的 1 年后,Sugar 达到了 10.4 kg(23 lb)的理想体重(图 9.7)。按之前同样的方式拍了照片,并加入病历记录中(图 9.6b)。我们将这些照片打印到证书上,并连同新的项圈一起赠送给 Sugar 的主人。在减肥计划中,Sugar 总共减肥 7.9 kg(17.4 lb)。为了鼓励维持体重,在现有减肥食谱基础上,给 Sugar 增加了 5%(17 kcal,每天总热量 353 kcal)的热量摄入。如有必要,可以每月复查 1 次以监督保持体重并调整热量摄入量。

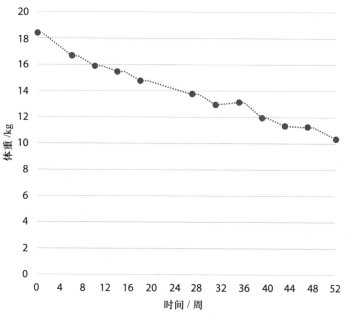

图 9.7　12 个月的减肥记录

9.8.6　讨论

兽医护士能在宠物的减肥计划中发挥重要的作用。通过整理收集包括正餐、零食、给药食物和营养补充剂在内的精确的饮食史，护士的工作为兽医制定饮食建议提供了强有力的基础。护士在宠物的减肥计划中发挥着积极的作用，并不断鼓励主人，这有助于对减肥计划的执行及克服过程中遇到的困难和压力。在候诊区放置记录宠物减肥过程的侧面照、每月的体重及减肥目标的展示板，也会起到激励主人的作用，并能产生良性竞争的效果，以及让主人互相了解到并不是只有他们遇到减肥平台期。Facebook 和其他社交媒体上的减肥参与者专区能给主人们提供倾诉和互相鼓励的平台，兽医团队也可以通过这种方式对主人进行宠物肥胖方面的指导。

当主人遇到一些减肥常见问题而感到挫败时，兽医护士能起到安慰和提供管理策略的作用。其中常见的问题就是主人要面对宠物不断的乞食行为。很多的喂食玩具能增加宠物进食过程的灵活性和复杂性。这类喂食玩具能延长进食时间并增加运动量。如果主人发现宠物对喂食玩具失去兴趣，换个喂食环境可能会减少宠物的乞食行为。举例来说，如果主人经常在厨房喂宠物并且经常出入厨房，宠物就会觉得主人一进厨房就可能有吃的。这时就可以换一个不常用来喂食的房间，比如客浴或者卧室。比较理想的是使用自动喂食器，这样有助于宠物将乞食对象从主人转移到喂食机器。

在一次复查结束前预约下次护士复查是主人的一种承诺。临近预约时,对主人的提醒也有助于宠主按时遵守预约或者根据情况改约其他时间。为了激励客户,可以在首次就诊时提供一个"启动包",建议放在"启动包"里的物品可以包括处方粮的试吃装、量杯、罐头盖、牵引绳,以及一些读本,比如减肥成功案例、每月互助会的传单、对体况评分的介绍及肥胖的危害等。每月互助会的话题可以多种多样,比如减肥宠物可以进行的运动、主人的成功故事、关于营养的问答环节,或者是在室内康复泳池的宠物团体游泳训练。如果管理得当,减肥计划可能会对宠物、主人和兽医诊所都产生积极影响。兽医护士通过训练学习如何指导客户、获取详细饮食史、掌控减肥过程、与主人讨论遇到的问题并给予解答,这也可以促进最终的减肥成功。首次就诊时护士主要辅助医生的检查工作,接受训练后,护士可以从医生的角度掌控之后的减肥复查。

当宠物达到了预估的理想/目标体重后,兽医师和护士会再次复查,并最终确定长期体重保持计划。总体来说,能够按减肥计划完成每次的复诊需要所有人共同努力。在整个减肥计划实行过程中,通过与兽医师的配合,兽医护士是指导和支持主人和他们的宠物成功完成挑战的最佳资源。

> **学习要点**
>
> (1)应对兽医护士进行培训,使他们能够轻松地检查和收集准确、完整的饮食史。饮食史的准确性会影响对于制定减肥计划时所用热量的估算。不够精确的热量估算会在计划启动后,导致不合理的体重变化。
>
> (2)宠物的减肥常常会进入平台期!因此,在这段时间,通过鼓励宠主对食物配方进行创造性改良,以及在互助会和社交媒体上的互动,对主人进行有效的指导和激励,有助于他们继续参与计划,渡过难关。
>
> (3)万物适度。零食可以是主人与宠物建立联系的一种方式,无须为了成功减肥而将其排除在饮食计划之外。要让主人认识到零食的热量分配在维持总体的能量平衡和达成减肥目标上的重要性。
>
> (4)找到宠物主人喂食过程中不可更改的地方非常重要。在回顾饮食史时,要确定零食习惯。比如,主人有睡前读书的习惯,而看书时他们会喂零食让宠物放松。这对主人和宠物都很重要,尽量提供适合能量分配的零食会在整体上影响宠物能否严格遵守减肥计划。
>
> (5)营养补充剂会提供热量。一定要考虑到零食中来自营养补充剂的热量。

参考文献

1. Brooks D, Churchill J, Fein K et al. 2014 AAHA weight management guidelines for dogs and cats. *Journal of the American Animal Hospital Association* 2014;50:1–11.
2. German AJ, Holden SL, Mason SL et al. Imprecision when using measuring cups to weigh out extruded dry kibbled food. *Journal of Animal Physiology and Animal Nutrition* 2011;95:368–373. https://www.ncbi.nlm.nih.gov/pubmed/21039926

索引

……的内分泌功能　endocrine functions of　19
……与肥胖症　and obesity　1, 6~7, 15, 19, 21~22, 36, 41, 78, 105
6 min 步行测试　6-minute walk test(6MWT)　43
ob/ob 小鼠　*ob/ob* rodents　19
Wishnofsky 法则　Wishnofsky's rule　108
阿黑皮素原（POMC）　Proopiomelanocortin (POMC)　16, 35
癌症　Cancer　20, 23, 34, 42~43, 46
巴吉特猎犬　Basset hounds　4
白细胞介素 –6　Interleukin-6 (IL-6)　41
半胱氨酸蛋白酶抑制剂 C　Cystatin C　40
胞外固体　Extracellular solids (ECS)　56
保持长期成功　long-term success　88
悖论　paradox　34, 38, 43, 46
比格犬　Beagles　6, 38, 44, 59, 82, 84~85
比较的案例　comparative example　175
必需营养素　Essential nutrients　83
变异系数　Coefficient of variation (CV)　58~59, 64
病毒　Viruses　15
病情评估　patient assessment　142, 148, 156
草酸钙　Calcium oxalate (CaOx)　40
策略　strategies　22, 154, 172, 177
产热　Thermogenesis　7, 15, 18
产热理论　"Thermogenic" theory　15
肠道微生物菌群　Gut microflora　21~22
肠道微生物菌群与肥胖症　gut microbiota and obesity　21
肠道微生物菌群在肥胖症治疗中的应用　utility of gut microbiota in obesity management　22
肠道相关淋巴组织　Gut associated lymphoid tissue (GALT)　171

常量营养素　macronutrients　6, 84
成本效益分析　Cost-benefit analysis　127~128
成年雌性史宾格猎犬的病态肥胖　Morbid obesity in adult female spayed Springer Spaniel　147
充血性心力衰竭　Congestive heart failure (CHF)　34, 38
宠物肥胖症　Pet obesity　4, 7, 8, 55, 65~66, 73, 95, 97, 105, 109,
宠物和主人关系　Pet-owner relationship　99
宠物行为　pet behavior　94
宠物与主人共同减肥的互动行为　Pairing pet and owner behavior for mutual success　101
宠物主人对宠物行为和肥胖症的主观认知　Owner perception on behavior and obesity　94
宠物主人主观认知对宠物行为和肥胖症的影响　Integrating psychology and behavioral management into obesity treatment　94
初次会诊　initial consultation　115, 116, 118, 122
处方粮　Therapeutic diets　83~86, 107~108, 117, 120, 143, 145, 147~151, 157, 160, 162, 164~165, 171, 178
垂体依赖性肾上腺机能亢进　Pituitary-dependent hyperadrenocorticism (PDH)　37
刺鼠相关蛋白　Agouti-related protein (AgRP)　16
促进日常运动　promoting daily exercise　37
促食神经肽　Orexigenic neurons　16
促炎性细胞因子　Pro-inflammatory cytokines　38
存储　storage　16, 18
达克斯猎犬　Dachshunds　4
代谢表现型　Metabolic phenotypes　33
代谢调控　Metabolic control　18
代谢和消除　metabolism and elimination　45
代谢能　Metabolizable energy (ME)　85, 142, 157, 171
代谢体重　Metabolic body weight (MBW)　107
单核细胞趋化蛋白-1　Monocyte chemotactic protein-1 (MCP-1)　41
胆固醇酯转运蛋白酶　Cholesterol ester transfer protein (CETP)　39
胆囊收缩素　Cholecystokinin (CCK)　17
蛋白质　Protein　6, 15, 17~19, 36, 40, 45, 56, 84~86, 143, 145, 157~158, 160, 162~163, 166~172
蛋白质饮食试验　Protein diet trial　166~167, 171

导致肥胖的行为　Obesogenic behaviors　99
低密度脂蛋白　Low-density lipoproteins (LDL)　39
低能量　Low-energy　6, 38, 84~87, 89, 145, 171
低热量水果和蔬菜饮食限量表　low calorie fruits and vegetables for treat allowance　87
低碳水化合物　Low carbohydrate (LC)　6, 36, 86, 162~163
低碳水化合物饮食　Low carbohydrate (LC) diets　86, 162
低纤维　low-fiber　36, 85, 162
低纤维饮食　Low-fiber diet　85
低氧血症　Hypoxemia　44, 151, 153~155
调整后的体重　Adjusted body weight (ABW)　152, 155
丁丙诺啡　Buprenorphine　45, 153
定量核磁共振 (QMR)　Quantitative magnetic resonance (QMR)　59~60
动脉收缩压　Systolic arterial blood pressures (SAP)　44
动脉血压　Arterial blood pressure　40
动脉血氧分压　Partial pressure of arterial oxygen (PaO$_2$)　44
动物自身因素　Animal-specific factors, 2; *see also Human-specific factors*　4~5, 8, 144
胴体分析　Body composition carcass analysis of　55~56, 58, 61
短期和长期随访　short-and long-term follow-up　143, 153, 161, 165
短期随访　short-term follow-up　157
对肥胖症的认知　perception of obesity　95
多不饱和脂肪酸　Polyunsaturated fatty acids (PUFAs)　171~172
多酚　Polyphenols　22
多能干细胞　Multipotent stromal stem cells(MSCs)　20
多尿和多饮　Polyuria and polydipsia (PU/PD)　160
多态性　Polymorphism　36
儿茶酚胺　Catecholamines　38
非处方日粮　Over-the-counter diets（OTC diets）　83
非饮食策略　Nondietary strategies　22
非影像学方法评估　nonimaging-based modalities to assess　60
非影像学方法评估体成分　Nonimaging-based modalitiesto assessing body composition　60
肥胖宠物评估　patient evaluation　73
肥胖的风险　and obesity risk　94~95
肥胖对心血管的影响　Cardiovascular effects of obesity　106
肥胖小动物的麻醉问题　Anesthetic concerns in obese small animal patient　151

肥胖症宠物的能量管理策略　Nutritional management of obesity energy requirements for weight management　78

肥胖症宠物的行为管理策略　Behavioral management of obesity creating effective weight management plans　94

肥胖症的进化理论　Evolutionary theories of obesity　14

肥胖症的炎症效应　Inflammatory effects of obesity　19

肥胖症发展过程　Develop obesity　15

肥胖症和寿命　Life span, obesity and　33

肥胖症和寿命　obesity and life span　33

肥胖症和炎症　obesity and inflammation　20

肥胖症流行　obesity epidemic　7~8

肥胖症麻醉的病理生理学考虑　Pathophysiology of obesity anesthetic considerations　33

肥胖症所致的药理学差异　Pharmacological differences in obesity　45

肥胖症引起的生理变化　Obesity-induced alterations in physiology　43

肥胖症治疗　in obesity management　22, 65, 97, 105, 107, 110, 166

分解代谢　Catabolism　18, 34, 38

分泌型抗体　Secretory antibodies　171

分配　distribution　45, 87, 98, 101, 118~119, 137, 156~158, 175, 178

粪菌移植　Transfaunation　22

风险因素　Risk factors　4~5, 8, 94~95, 102, 142

腹部膨胀　Abdominal distension　37

甘油三酯　Triglycerides　20, 37~38, 56

感染　Infections　40, 43, 153, 166, 169, 171

高蛋白　Elevated protein　15, 36, 84~86, 162~163, 171~172

高蛋白质高纤维饮食（HPHF）　High protein-high fiber diet (HPHF)　85

高甘油三酸酯血症　Hypertriglyceridemia　160

高密度脂蛋白　High-density lipoproteins (HDL)　39

高碳酸血症　Hypercapnia　43~44

高纤维　High fiber (HF)　36, 85~86, 97, 162~163

高血糖症　Hyperglycemia　160

高血压　Hypertension　19, 39, 44~45

高脂血症　Hyperlipidemia　37~38

个性化减肥计划　Individualized weight loss plan　145

跟进和复查预约　follow-up and recheck appointments　17

功能余气量　Functional residual capacity(FRC)　43
估算理想体重的方程　equations to estimating ideal　77
估算目标或理想（体重）　estimating target or ideal　77
骨关节炎　Osteoarthritis（OA）　41, 43, 45, 65, 105, 107~108, 110
骨科和神经内科疾病　orthopedic and neurologic disease　41
骨科疾病　Orthopedic disease　41~42
骨肿瘤　Bone tumors　42~43
观念　perception　8, 95, 171
管理处方粮　management diet　83~84
过度喂食　Overfeeding　7, 87
过量活性氧 (ROS)　Reactive oxygen species (ROS)　21
过敏　hypersensitivity　166~172
哈巴犬　Pugs　4
含有能量的营养物质　energy-containing nutrients　17
合并症　Comorbidities　17, 33~35, 37, 39, 41~43, 45, 73, 77, 83, 97~98, 105~107, 110
合并症、运动和肥胖症　comorbidities, exercise, and obesity in　107
合成代谢　Anabolism　18
核磁共振成像 (MRI)　Magnetic resonance imaging (MRI)　59~60, 62, 65
黑皮质素受体 4　Melanocortin 4 receptor (MC4R)　36
恒脂器　Adipostat　17
厚壁菌门成员　Firmicutes phylum members　21
呼吸急促　Hyperpnea　107
呼吸系统疾病　Respiratory disease　38, 43
呼吸系统疾病病理生理学　respiratory pathophysiology　43
环境因素　Environmental factors　15
环氧合酶 2　Cyclooxygenase 2 (COX2)　21
患病率和时间趋势　prevalence and time trends　1
患骨关节炎的超重犬　Osteoarthritic overweight dogs　108
回归分析　Regression analysis　33
肌肉状况评分　Muscle condition score (MCS)　65, 78~79, 88, 142, 167
基础代谢　Basal metabolism　5, 18~19
基础代谢率　Basal metabolic rate (BMR)　5, 18~19
基础全血细胞计数　Complete blood count (CBC)　139
基于人类情感的案例　emotional example　98

基于信息交流的案例　Information-based example　97
激素　hormones　6, 16, 18, 21, 34, 37, 42, 107, 167
极低密度脂蛋白　Very low-density lipoproteins (VLDL)　39
计算机断层扫描　Computer tomography (CT)　56, 58
计算每周减肥率　calculating rate of weight loss per week　88
既往病史　patient history　142, 147, 151, 156, 160, 166, 174
家庭成员互动　family dynamics in　100
家庭成员互动在管理肥胖行为　Family dynamics in behavioral management of obesity　100
家庭特征　Household characteristics　4, 7~8
家养短毛猫　Domestic shorthair cat (DSH cat)　36, 151~152, 156, 160
甲氧苄啶-磺胺嘧啶　Trimethoprim-sulfadiazine，（TMS）　167
甲状腺　Thyroid gland　18, 23, 37, 167
甲状腺功能减退　hypothyroidism　23, 37, 167
甲状腺激素　Thyroid hormones　18, 37
甲状腺素　Thyroxine (T4)　157, 164, 175
减肥并保持体重　Weight loss and weight maintenance;see also Overweight　117
减肥处方粮　weight loss diets　147, 157
减肥计划中的肥胖程度估算　Degree of obesity estimation for weight loss planning　156
减肥记录　body weight loss history　177
减肥食物分配　Portioning food for weight loss　87
建议用量　Recommended allowances (RA)　172
结肠切除术　Subtotal colectomy　151~152
解决情绪行为　Addressing emotional behaviors　99
解偶联蛋白1基因(UCP1基因)　Uncoupling protein 1 gene (UCP1 gene)　15
金毛猎犬　Golden retrievers　4
进食动力　food motivation　165
经济拮据　Financial constraints　126
经济因素的影响　Financial implications　126
静息能量需要　Resting energy requirement (RER)　18, 81~83, 148, 156~158, 161, 168, 175
局部炎症　Local inflammation　20
聚集素　Clusterin　40
凯恩狸　Cairn terriers　4

抗生素　Antibiotics　22, 170

抗炎激素脂联素　Anti-inflammatory hormone adiponectin　21

可卡犬　Cocker spaniels　4

可卡因-苯丙胺调节转录物　Cocaine- and amphetamine-regulated transcript (CART)　16

客户沟通　Client communication　127, 145, 166

客户照片发布许可表　Client photograph release consent form　136

腊肠犬中　in dachshund　42

了解宠物主人的想法　understanding owner perceptions　95

临床环境体成分的评估　estimating body composition in a clinical setting　62

临床环境中的评估　estimating in clinical setting　62

淋巴管浸润　Lymphatic invasion　42

流行病学研究　Epidemiologic study　5, 19, 40, 94, 105

麻醉注意事项　Anesthetic considerations　43

马拉色菌　*Malassezia*　166~167, 170

慢性低度炎症状态　Chronic low-grade inflammation　36

慢性疾病　Chronic illnesses　20, 23, 34

慢性肾脏病　Chronic kidney disease (CKD)　34~35, 46

慢性炎症状态　Chronic inflammatory states　19, 21

猫的强化训练　Enrichment program in cats　108

猫肥胖症的患病率　prevalence of obesity in canine populations　1, 3

猫科动物　Feline　75~76, 107, 126

猫咪（合并症、运动和肥胖症）　Cats (comorbidities, exercise, and obesity in)　123

猫咪之夜　"Cat's Night Out" events　125

猫糖尿病　feline DM　36, 107

猫特发性膀胱炎　Feline idiopathic cystitis (FIC)　40

猫体重指数　Feline body mass index (FBMI)　63

猫下泌尿道综合征　Feline lower urinary tract disease(FLUTD)　40

每日蛋白质摄入量　Daily protein intake　84

每周饮食记录　Weekly food diary　139

美国动物医院协会　American Animal Hospital Association (AAHA)　55, 156, 165

美国农业部　United States Department of Agriculture (USDA)　81, 149

美国兽医运动医学与康复学院认证医师　Diplomate of the American College of Veterinary Sports Medicine and Rehabilitation (DACVSMR)　126

美国饲料管理协会　Association of American Feed Control Officials (AAFCO)　81, 145

美国养犬俱乐部　American Kennel Club (AKC)　33
觅食行为　food-seeking behaviors　17
泌尿疾病　Urinary disease　39~40
密度　density　6, 38~39, 59, 84~86, 89, 145, 162, 171
免疫　Immunotherapy　19, 21, 167, 170
目标体重　Target body weight (TBW)　46, 58~60, 73, 77, 81, 83~84, 89, 106, 116, 119~121, 152, 161, 171, 176, 178
目标体重维持　Target weight maintenance process　121
内分泌疾病　Endocrine disease　35, 37, 39, 45
内脏肥胖　Visceral obesity　37
内脏脂肪　Visceral adiposity　37, 59
能量摄入　Energy intake　1, 4, 6~8, 14, 17~18, 22, 33, 36, 38, 73, 78, 81~86, 89, 105, 107~109, 115, 142~145, 150, 159, 168, 176
能量消耗　Energy expenditure　5~7, 14, 16, 18~19, 37, 83~85, 108~110, 145, 169~171
能量消耗和能量代谢调控机制　mechanisms of energy expenditure and metabolic control　18
能量消耗与能量摄入的平衡　Balancing energy expenditure and intake　14
能量正平衡　Positive energy balance　1, 4
拟人化的行为　Anthropomorphic behaviors　98
年龄　Age　5, 8, 41, 78, 94, 116, 129, 131, 158, 161, 167, 169, 171
尿白蛋白肌酐比值　Urine albumin:creatinine ratio (UAC ratio)　40
尿蛋白肌酐比值　Urine protein:creatinine ratio (UPC ratio)　40
尿道梗阻　Urethral obstruction (UO)　40
尿路感染　Urolithiasis, urinary tract infections (UTI)　40
欧洲宠物食品工业联合会　Fédération européenne de l'industrie des aliments pour animaux familiers (FEDIAF)　81
陪伴　companionship　58~60, 62, 65
漂变基因　Drifty genes　14
品种　Breed　4~5, 14, 33, 62~65, 106, 116, 129, 131, 143, 145, 148, 158, 171
评分系统　scoring system　33~34, 39, 44, 63, 65, 74, 79, 116, 158
评价方法和术语　measurements and terms　57
葡萄糖代谢　Glucose metabolism　20, 36
气道阻力　Airway resistance　43~44, 107
气管塌陷　Collapsing trachea　39, 45

器官功能障碍　organ dysfunction　45
牵张感受器　Stretch receptors　16~17
轻度的高白蛋白血症　Mild hyperalbuminemia　160
氢核　Hydrogen nuclei　59
情感依赖　Emotional dependency　99
去势/阉割　Neutering　82
去脂体重　Fat-free mass (FFM)　56~58, 60~61, 65, 171
全科诊疗的临床建议　clinical recommendations for general practice　65
全身炎症　Systemic inflammation　21, 44, 107
犬肥胖症的患病率　prevalence of obesity in feline population　1
犬科动物　Canine　75
犬猫……并发症　Comorbidities … in dogs and in cats　155
犬糖尿病　canine DM　35
人类肥胖症管理　in human obesity management　106
人类与宠物关系　Human-animal relationships　102
人为因素　Human-specific factors　4~6, 8, 144
日粮　Diet(ary)　45, 87, 98, 101, 118~119, 137, 156~158, 175, 178
日粮类型和喂养方式　diet type and feeding method　6
日粮诱导肥胖大鼠　diet-induced obese rats　41
融入心理/结合心理学/到心理学　integrating psychology and　166
乳酸杆菌　*Lactobacillus*　22
商业宠物食品类型　Commercial diet types　6
上气道功能障碍　Upper airway dysfunction　44
设得兰牧羊犬　Shetland sheepdogs　4
摄入量　intake　6, 36~38, 73, 78, 81~88, 108~110, 115, 119~121, 142~145, 167~170, 176
身体康复　Physical rehabilitation　147, 149
神经功能障碍　Neurologic dysfunction　42
神经内科疾病　Neurologic disease　41~42
神经肽 Y(NPY)　Neuropeptide Y (NPY)　16
神经元　Neurons　16, 20
肾上腺皮质功能亢进　Hyperadrenocorticism (HAC)　37
肾素-血管紧张素系统　Renin-angiotensin system　21, 40
肾小球超滤　Glomerular hyperfiltration　40
肾脏和泌尿疾病　renal and urinary disease　39

索引

肾脏疾病　Renal disease　38~39, 77

生活环境　Habitat　7, 21, 95, 102, 134~135

生物电阻抗　Bioimpedance (BIA)　56, 58, 61, 65

时间趋势　Time trends　1, 4

实施减肥计划　implementing weight loss plan　8, 43, 81, 83

实施治疗方案　implementation of treatment program　84

实用建议　practical recommendations　109

食物热效应 (TEF)　Thermic effect of food (TEF)　18

食物选择　Diet selection　17, 84

食欲的快感（享乐）调控　Hedonic (pleasure) control of appetite　17

食欲的外周调节　Peripheral regulation of appetite　16

食欲的中枢调节　Central regulation of appetite　16

食欲调节　Appetite regulation　21, 36

食欲和采食的调节　Regulation of appetite and food intake　16

世界小动物兽医协会　World Small Animal Veterinary Association (WSAVA)　55, 74, 78~79, 116, 142

寿命　Life span　5, 33, 40, 46, 65, 122

受到特殊关注的微恙猫咪健身房　special focus on Bug's Cat Gym　122

兽医护士　Veterinary nurses　116, 118~121, 127~128, 174~175, 177~178

兽医教学医院　Veterinary teaching hospital (VTH)　147, 156, 174

兽医学　Veterinary medicine　34, 40, 97, 129, 136

兽医诊疗　Veterinary practices　149

兽医制定饮食　Veterinary therapeutic diet　177

瘦素　Leptin　16~17, 19, 21, 36~38

瘦体重　Lean body mass　45, 56~57, 60, 63, 77~78, 84, 108~109, 143, 145, 152, 155, 162, 167, 170

舒张功能障碍　Diastolic dysfunction　44

双侧颅内交叉韧带　Cranial cruciate ligament (CCL)　41

双能 X 射线吸收法　Dual energy x-ray absorptiometry(DEXA)　56

双歧杆菌　*Bifidobacterium*　22

双歧杆菌　*Enterococcus*　22

水解后的食物　Hydrolyzed diets　171

水下跑步机疗法　Underwater treadmill therapy　110

饲喂　Feeding　36, 84~89, 94, 117, 121, 132~133, 137~138, 147~149, 157~158,

161~163, 174

随访　follow-up　83, 88~89, 106, 115, 118~119, 121, 127, 143, 153, 157, 161, 165, 169~170, 176

肽YY（PYY）　Peptide YY（PYY）　17

碳水化合物　Carbohydrates　6, 15, 18, 36, 86, 157, 160, 162~163

碳水化合物日粮　Dietary carbohydrate　6, 15, 36

糖尿病　Diabetes mellitus (DM)　15, 19, 23, 35~38, 45, 65, 86, 98, 106~107, 160~163

讨论宠物肥胖问题　discussion of pet obesity　95

体成分　Body composition　1, 5, 41~42, 45, 55~63, 65, 73

体成分的胴体分析　Carcass analysis of body composition　55

体况评分　Body condition score (BCS)　2, 4~5, 33~34, 36~42, 44, 55~56, 58, 62~65, 73~75, 77~78, 86, 88, 95~96, 116, 119, 121, 128, 142~143, 147~148, 156, 159~162, 164, 167, 170, 175, 178

体细胞质量　Body cell mass　56

体脂　Body fat　6, 14, 17, 19~20, 40, 45, 56, 58~59, 61~62, 64~66, 73, 86, 121, 142, 158, 160~161

体脂率　Body fat percentage (BF%)　58, 61~62, 64~66, 73, 75, 77~78, 157~158, 160, 175

体脂率（BF）的形态计量预测方程　predictive equation for body fat percentage using　77

体脂率及超重　with body fat percentage and excess body weight　75

体重超标　Excess body weight　1, 99, 101

体重管理的能量需要　energy requirements for weight management　78

体重指数　Body mass index (BMI)　8, 14, 61~63, 153

同型半胱氨酸　Homocysteine　40

退变性关节病　Degenerative joint disease (DJD)　41

微生物在肥胖症中的作用　Microbiome role in obesity　21

微恙猫咪健身房　Bug's Cat Gym　122~123, 125

微恙猫咪攀爬山　Bug's Mountain　122, 124

维持计划　maintenance appointments　144

维持能量需要　Maintenance energy requirement (MER)　18, 81, 89, 148, 161, 168

委员会认证的兽医营养师　Board Certified Veterinary Nutritionist™　83

胃肠道（GI tract）　Gastrointestinal tract (GI tract)　21, 43, 56, 145

喂食行为　Feeding behaviors　95, 97, 99~100, 176

物理治疗　Physiotherapy　108

吸收　absorption　21, 40, 44~45, 56~58, 60, 157

细胞色素 p450　Cytochrome p450 (CYP)　46
细菌　Bacteria　15, 21, 166~167, 170
下丘脑　Hypothalamus　16~18
限制食物摄入　Substantial food restriction　6
消耗　expenditure　6, 14~15, 18, 23, 84, 85~87, 89, 105, 107~108, 110, 117, 139, 150, 167, 170
小贴士　quick tips　118, 120~121, 126, 128
心理学角度　Psychological perspective　94, 166
心血管病理生理学　Cardiovascular pathophysiology　44
心血管和呼吸系统疾病　cardiovascular and respiratory disease　38
心血管疾病　Cardiovascular disease　19, 21, 38, 77
新生宠物管理　Neonatal management　22
新型蛋白质饮食试验　novel protein diet trial　166
新型蛋白质饮食试验中的热量限制　Caloric restriction … during novel protein diet trial　166
新形态方程　Novel morphometric equations　63
行为风险因素　Behavioral risk factors　95
行为管理　behavioral management of　94, 164~166
形态学估算　Morphologic estimates　63
性别　Gender　5, 62, 82, 116, 129, 131, 171
性别和绝育　gender and neutering　5
血浆药物清除率　Plasma clearance　46
血清 IgE 抗体　Serum IgE antibodies　167, 171
血脂异常　Dyslipidemia　38, 106
炎性细胞因子　Inflammatory cytokine　20~21, 38
炎症　Inflammation　14, 19~21, 36, 41, 44, 78, 107
厌食神经元　Anorexigenic neurons　16
药代动力学研究　Pharmacokinetic studies　46
药物　Drug　17, 43, 45~46, 115, 127, 130, 133, 147, 152~155, 162, 170
药物代谢与消除　drug metabolism and elimination　46
药物分布　drug distribution　45
药物吸收　drug absorption　45
一例新型蛋白质饮食试验中的治疗性减肥　Therapeutic weight loss during novel protein diet trial　166

一氧化氮　Nitric oxide　21

医学因素　Medical perspective　113

胰岛素抵抗　Insulin resistance　17, 19~20, 35~36, 38, 65, 86, 106, 162

胰岛素增敏剂　Insulin sensitizer　19

胰高血糖素样肽-1　Glucagon-like peptide-1 (GLP-1)　17

遗传　Genetics　4, 14~15, 107

遗传和品种　genetics and breed　4

遗传因素　Genetic factors　4, 14

以家庭为导向的肥胖治疗项目　Family-oriented obesity treatment programs　102

易感品种　Breed predisposition　161

益生菌　Probiotics　22

饮食　diet　6, 8, 15, 17, 21~22, 36~38, 85~86, 115

饮食、活动和家庭生活史表　Diet, activity, and household history form (DAHHF)　78

饮食策略　dietary strategies　22, 171

饮食和运动　diet and exercise　94~95, 172

饮食历史　diet history　83, 115~116, 118, 150

饮食限额表　Treat allowance sheets　87, 117~118, 128, 136

饮食形式和能量密度　diet form and energy density　86

营养评价　Nutritional assessment　55, 142, 144

营养失调　Nutritional disorder　55, 94, 166

影响基础代谢　affects basal metabolism　18

影像学方法评估体成分　Imaging modalities to assessing body composition　57

"优质"干粮　"Premium" dry foods　6

由兽医指导的运动干预　Veterinary-led exercise interventions　108

游离脂肪酸　Free fatty acids (FFA)　20

有效的体重管理计划　Effective weight management plans　99

有益健康的行为　Fostering healthy behaviors　102

与宠物主人相关的导致宠物肥胖的危险因素　owner-related risk factors for pet obesity　94

预吸氧　Preoxygenation　44, 152~154

运动　Exercise　7~8, 15, 19, 37, 39, 41~42, 44, 62, 64, 73, 94~95, 100~101, 105~110, 116~117, 122, 125~126, 134, 139, 144, 149~150, 164~172, 177~178

运动和肥胖症　exercise and obesity risk　107

运动减少　Reduced exercise　5, 105

运动量　Physical activity　4~5, 7~8, 94, 105, 106~109, 139, 144, 150, 170, 177

运动与人类肥胖症管理　exercise in human obesity management　106

运动在犬和猫肥胖症治疗中的应用　exercise in canine and feline obesity protocols　107

肥胖的短毛猫　in obese DSH cat　156

在最初的体重讨论中参考兽医的意见　incorporating perception in initial weight discussion　99

增生　Hyperplasia　20

长期正能量平衡　Chronic positive energy balance　1, 4

诊断　diagnosis　8, 34, 39, 42, 55, 73, 156, 160, 162, 167, 171

拯救宠物减肥大赛　Pets Reducing for Rescues contest (PRFR contest)　125

正呼气末压　PEEP valve　153, 154

正交偏最小二乘判别分析　Orthogonal projections to latent structures discriminant analysis（OPLS-DA）　33

脂蛋白代谢　Lipoprotein metabolism　33

脂肪　Fat, 8; *see also* Obesity　1~3, 6, 14~23, 34~39, 41~46, 55~61, 63~65, 77~78, 86, 106, 117, 142~143, 145, 151~153, 156~158, 160, 162

脂肪代谢　Lipid metabolism　38

脂肪合成　Lipogenesis　21

脂肪-水 MRI　Fat-water MRI (FWMRI)　59

脂肪细胞　Adipocytes　16, 18~21, 42

脂肪因子　Adipokines　19~21, 34, 36~37, 106

脂肪组织　Adipose tissue　14~23, 34~36, 39, 41, 45~46, 56, 58~61

脂肪组织的内分泌功能　Endocrine functions of adipose tissue　19

脂联素　Adiponectin　19~21, 36~37, 42, 106

指数与线性方程　exponential *vs.* linear equation　83

制定饮食指南　setting treat guidelines　87

治疗方案　plan　73, 84, 94~95, 97, 102, 117, 143, 148~149, 151, 156, 160, 164, 167~168, 175

治疗肥胖对于缓解糖尿病　Diabetes mellitus remission for weight loss　162

治疗性减肥　Therapeutic weight　166, 175

中蛋白质高纤维饮食　Moderate protein-high fiber diet(MPHF)　85

中枢神经系统　Central nervous system　16

肿瘤　Neoplasia　20, 40, 42~43, 46

肿瘤坏死因子 –α　Tumor necrosis factor-α, (TNF-α)　19, 36

重水（D_2O）稀释技术　D_2O dilution technique　59~61
重新评估减肥计划　Reevaluating weight loss program　88, 145
主人态度和家庭特征　owner attitudes and household characteristics　4, 7~8
椎间盘疾病　Intervertebral disc disease (IVDD)　42
自由　*Ad libitum*　6~7, 17, 36, 60, 85, 89, 105, 156
棕色脂肪组织　Brown adipose tissue (BAT)　15, 18
总膳食纤维　Total dietary fiber (TDF)　85, 145
组织　mass　5, 14~23, 34~36, 38~42, 45~46, 55~61, 63, 84, 86, 171
组织血流量　Tissue blood flow　45